SIDONIUS APOLLINARIS

AND HIS AGE

OXFORD
UNIVERSITY PRESS
AMEN HOUSE, E.C. 4
London Edinburgh Glasgow
Leipzig New York Toronto
Melbourne Capetown Bombay
Calcutta Madras Shanghai
HUMPHREY MILFORD
PUBLISHER TO THE
UNIVERSITY

SIDONIUS
APOLLINARIS
AND HIS AGE

By

C. E. STEVENS
B.A., B.Litt.

Formerly Scholar of New College and Robinson
Exhibitioner of Oriel College

OXFORD

AT THE CLARENDON PRESS

1933

PRINTED IN GREAT BRITAIN AT THE UNIVERSITY PRESS, OXFORD
BY JOHN JOHNSON, PRINTER TO THE UNIVERSITY

ONULP

PREFACE

MEN have often wondered whether we could say of a single date that it marks the end of the Old World and the beginning of the Middle Ages. A.D. 476 was often taken—and very naturally—as the transition point. But Mommsen thought that the New Empire of Diocletian marked the change of an epoch; others have pointed to the conversion of Constantine or the founding of Constantinople as the decisive date; the student of art, again, remembering the 'Dijon school' of Gallo-Roman sculpture or the bust of Magnentius at Vienne, may hold different views. It was a service of nineteenth-century historians to point out that medieval institutions have their roots deep in the Roman Empire and that A.D. 476 was not to be regarded as a great thunder-clap with which a new world was born. Perhaps in this they exaggerated 'the inevitability of gradualness' and forgot that epochs of swift transition do indeed occur. But we should not forget it, who are living now in such a one. For us Sidonius should have a particular interest inasmuch as he justifies us in regarding the second half of the fifth century as one of those periods. We see in him the last product of the Classical Education, the last representative of the educated Roman aristocracy. He lamented that his age should see the failure of an epoch, and his lament was justified. Learning fled to the cloister, where it remained for nearly a thousand years, and thus the age of Sidonius presents to us a break from the past of the most momentous consequences for the future.

In another aspect of thought we can see the transition actually working in the life of Sidonius himself. He

was born into an age when men still regarded the
Roman Empire and civilization as conterminous. If
a usurper arose in Gaul or Britain he did not consider
himself emperor of Gaul or Britain, but as at least a
partner in the administration of the whole. Though
one emperor ruled at Rome and another at Constantinople, the empire was regarded as one. Laws pro-
pounded by one emperor were automatically valid in
the dominions of his colleague, and if one emperor
died, the administration of the whole devolved upon
the survivor. Even if a barbarian king settled in a
province of the empire, he was regarded as king only
of his own tribe; the indigenous population were
still members of the empire and lived under Roman
laws. Regional patriotism existed, indeed, but regional
patriotism has never been inconsistent with a larger
loyalty. But this loyalty could not survive the disruptive forces of the age, and with Sidonius and his
contemporaries we look into the hearts of men who
saw it die. We see Arvandus demanding that Gaul
should be divided among barbarian kings 'according
to the law of nations'; we see Sidonius beaten at last
in the fight that he waged so gallantly in word and
deed for the maintenance of the old ideals, and we
can read the poem of praise in which he recognized the
claim of Euric to 'rule Gaul', as Jordanes said, 'in his
own right'. It was, indeed, a break from the past when
Euric could take for his own the very title 'Dominus
Noster', which had belonged to the imperial house; when
Clovis could issue his new and independent coinage.
These men were laying the first foundations of European
nationalism, and Sidonius is a witness of the work.

And no judge in a court of law could ask for a better

witness. He was a faithful observer of his age, and he did not distort his observations by a preconceived theory of their meaning. He had no philosophy of history; but that does not diminish his value to the historian, and he tells the story of these changes no less truly for not appreciating their significance.

He was an honest man, but not a great man; an actor, but not a principal actor on the stage of history; and thus we might in any event be more interested in his evidence than in himself. And, as it happens, such an attitude towards his works is almost forced upon us by those works themselves. Sidonius collected his Letters for publication when he was already forty-five years of age, and he revised and rewrote them with the greatest care. In this revision he smoothed out the contours of his life: we see not the development of a character, but a collection of formal pictures illustrating the manners of the fifth century. Around Sidonius moves the circle of his universe; it is a varied universe, but he stands at rest in the midst of it.

Thus a biography of Sidonius not only ought to be, but must be, a history of 'Sidonius and his Age', and for this reason it has not been thought necessary to include in the text any long extracts or paraphrases of his writings, Moreover, the Letters can be conveniently studied in O. M. Dalton's two volumes. It would be difficult to praise too highly the translation and the introduction that accompanies it. With Sirmond's edition it forms the essential avenue to an understanding of Sidonius' work. My debt to Dr. Dalton is very great, and I have particularly to thank him for permitting me to make use of his rendering in my quotations from the Letters.

But in whatever way the biography of Sidonius is treated, a particular difficulty faces the inquirer: the letters and poems are not arranged in any chronological order, and seldom contain in themselves clear evidence of date, while the other authorities for the period are both scanty and exceedingly complicated. To sort the documents into an order is a task entailing much detailed study such as few casual readers of Sidonius have either time or inclination to give. Moreover, it is a task in which complete success is unattainable; and I hope I shall not be accused of dealing ungenerously with my predecessors in the field, to whom my debt is great, when I say that they have made the chronology of Sidonius' life seem much more precise than it is. It is hard to blame them, for unless definite decisions are made in regard to the succession of events it would not be possible to write a biography of Sidonius at all. Definite decisions must indeed be made, but it is the historian's duty to explain why he makes them and what degree of probability he assigns to them. This I have endeavoured to do in every instance. Sometimes the arguments may appear slender: occasionally dates have had to be adopted arbitrarily when there is no evidence at all to allow of precision. Such cases are fortunately rare; more common are those where we can only balance probabilities with more or less hope of being right. Nevertheless, I trust that this chronological uncertainty will not prove so menacing in practice as might appear, and that most of the important dates may be regarded as reasonably sure.

It is a pleasant task to acknowledge the help of many scholars both abroad and at home. I wish first to thank MM. A. Audollent, E. Bréhier, and G. Desdevises du Dézert of Clermont-Ferrand University for their friendly hospitality and assistance; I wish also to acknowledge the assistance of M. Camille Jullian and M. R. Lantier. Of those in England who have helped me, I must first mention the Provost and Fellows of Oriel College: as Robinson Exhibitioner of the foundation, I was enabled to carry out my work. I have also particularly to thank Professor Baynes, Professor Jacob, R. G. Collingwood, I. A. Richmond, and J. N. L. Myres. Finally I would acknowledge my debt to the late Professor C. H. Turner, formerly Dean Ireland's Professor of Exegesis in the University of Oxford. His friends know well the eager encouragement that he was always so ready to bestow, and I hope that they will think it a fair tribute to his memory if I dedicate this work to him.

ST. GERMAIN-EN-LAYE,   C. E. S.
January 1933.

CONTENTS

ILLUSTRATIONS

CHRONOLOGY

A.D.	*Life of Sidonius*	*Roman Emperors*	*Visigothic Kings*
418			Theodoric I.
425		Valentinian III.	
432?	Birth (Lyons), Nov. 5th.		
451			Thorismund.
c. 452	Marries Papianilla.		
453			Theodoric II.
c. 454	First extant Letter (*E.* i. 2).		
455		Petronius Maximus.	
	Goes to Rome with Avitus.	Eparchius Avitus.	
456	Jan. 1st, Panegyricus Aviti.		
456?	Returns to Lyons.	Oct. Deposition of	
456–7	Coniuratio Marcelliana.	Avitus.	
457		June? Flavius Maiorianus.	
458	(Nov. or Dec.), Panegyricus Maioriani (Lyons).		
459	At Arles (cf. *E.* ix. 13. 4). Appointed Comes Arvernorum.		
461	At Arles again (*E.* i. 11).	Libius Severus.	
461–7	Private life (Lyons and Auvergne).		
465		Death of Severus.	
466		Anthemius.	Euric.
467	Sidonius at Rome.		
468	Jan. 1st, Panegyricus Anthemii. Publication of Official Poems. Appointed Praef. Urbi.		
469	Returns to Auvergne. Complete collection of Poems published. (Autumn) (?) Elected Bishop of Clermont.		
469–70	(Winter) (?) Journey to Ruteni.		
470	Conducts Episcopal election at Bourges.		
471–4	Visigothic invasions of Auvergne.		
472		Olybrius.	

A.D.	*Life of Sidonius*	*Roman Emperors*	*Visigothic Kings*
473		Glycerius.	
474		Julius Nepos.	
475	Auvergne surrendered to Visigoths.		
475		Augustulus.	
475–6	Sidonius confined at Capendu.		
476		End of Western Empire.	
476	At Bordeaux (*E.* viii. 9).	Odoacer.	
476?	Return to Clermont.		
477–80	Publication of Epistles. inter 480 et 490 Death (Clermont).		

ABBREVIATIONS

C.S.E.L. = *Corpus Scriptorum Ecclesiasticorum Latinorum.*

F.H.G. = *Fragmenta Historicorum Graecorum.*

M.G.H. = *Monumenta Germaniae Historica.*

P.-W. = Pauly-Wissowa, *Realencyklopädie der classischer Altertumswissenschaft.*

Patr. Graec. & Lat. = Migne's *Patrologia Graeca & Latina.*

I

YOUTH AND EDUCATION

C. SOLLIUS APOLLINARIS SIDONIUS,[1] to give him his full name, was born at Lyons[2] on November 5th, 432.[3] He was one of several children,[4] and belonged to an old family of the Lyonnais, one of those aristocratic houses[5] of Gaul, among whom public service was a tradition. His father and grandfather had both filled the office of praetorian prefect of Gaul, the most important dignity in the civil administration to which a subject could attain.[6] His mother, of whom he tells surprisingly little, was connected with the Aviti.[7] The son was thus an aristocrat by birth, and from his cradle a member of the Senatorial order. An aristocrat with the pride of ancestry behind him he remained. The picture of the saintly and in truth slightly priggish Sidonius which the biographers have drawn may lead one to forget that there

---

[1] See Mommsen, *ap.* Lütjohann, p. xlvi. The name Modestus occurs only in a few inferior manuscripts of his works: it may be a mistaken inference from *E.* ix. 12. 3.   [2] Coville, pp. 35–7.

[3] The day is given by *C.* xx. 1. For the year see *E.* viii. 6. 5, 'audiui eum (*sc.* Flauium Nicetium) adulescens atque adhuc nuper ex puero, cum pater meus . . . Gallicanis tribunalibus praesideret, sub cuius . . . magistratu consul Astyrius anni fores uotiuum trabeatus aperuerat.' Astyrius was consul in 449 (see Liebenam, *Fasti*, p. 46), and thus our date for Sidonius' birth will depend upon the precise meaning of 'adulescens atque adhuc nuper ex puero'. We learn from Gellius, x. 28 (cf. Jordanes, *Get.*, lv. 282) that *adulescentia* began at seventeen and from *Cod. Theod.*, ii. 17. 1, that it ended at twenty. Thus we have the limits of Nov. 429–Nov. 432, and 'nuper ex puero' suggests the latest date within these years.

[4] Sidonius had a brother, who seems to have been educated for the Church (*C.* xvi. 72–4). He is not elsewhere mentioned and may have died young. He had also more than one sister (*E.* v. 16. 5). Other members of the family: *E.* iii. 12. 5. (uncle), and *E.* i. 5. 2. ('sodalium propinquorumque').

[5] Cf. Greg. Tur., *Hist. Franc.*, ii. 15. (21), 'uir secundum saeculi dignitatem nobilissimus et de primis Galliarum senatoribus'.

[6] For the ancestry of Sidonius see *E.* i. 3. 1; i. 7. 7; iii. 12. 5, vv. 1–20; v. 9. 1, 2; viii. 6. 5; and Mommsen (*ap.* Lütjohann, p. xlvii).

[7] *E.* iii. 1. 1.

were influences which from his birth were to lay the mark of very different characteristics upon him. Even when he was bishop, he could pray that his son should become not a bishop but a consul.[1] 'Sidonius boasts', said his detractors at Bourges, 'of his birth; he looks down on us from the height of his official position, he despises the poor of Christ.'[2] Detractors are seldom entirely astray about a man's character.

There was another influence which must have contributed much to form his character in his earlier years —the state of his country at the time. Here was a boy born at a time when the picture of Gaul presented in the melancholy poem of Orientius was fresh in all men's minds. Only a few years before, 'Death, fire, and lamentation were seen in village and in farm, in country and in street, through every district. All Gaul smoked in one funeral pyre.'[3] And perhaps when he was already old enough to understand, men complained how not a castle on the rocks, not a town in the high mountains, not a city on the broadest river could withstand the fury of the barbarian.[4] When he was six 'nearly all the slaves of Gaul joined the Bacaudae', those bands of brigands that roamed about the countryside.[5] Four years later a troop of barbarians from the Caucasus were cantoned among the deserted fields of Valence[6] within sixty miles of his home. He was twenty-two when the Huns swept in their all-destroying march across north Gaul. Into the soul of such a man the iron should surely have entered, and in his correspondence we should expect to find clearly marked the shadow of

---

[1] *E.* v. 16. 4.   [2] *E.* vii. 9. 14.   [3] Orientius, *Commonitorium*, ii. 181–4.
[4] ps.-Prosper, *Carmen de Prouidentia Dei*, 35–8 (Migne, *Patr. Lat.*, li, p. 618).
[5] *Chron. Gall.*, a. ccclii, 117 (i, p. 660).   [6] Ib. 124 (i, p. 660).

his earlier days. Yet his most sympathetic interpreter can say, with perhaps an exaggeration but certainly not a perversion of the truth, that the life which Sidonius describes might be that of Hadrian's or Trajan's days.[1] Economically the land had made a recovery which must have needed a tough pertinacity and resolution, like that of the Frenchman of to-day.[2] The economic recovery that men's hands had effected, worked too upon their minds, and thus is explained that unexpected serenity which we notice in the letters of Sidonius.

There was another influence which could assist in diverting a young man's mind from the havoc of his time, for its aim was to turn it towards the past. That was the system of education under which he grew up. There are scattered among the letters of Sidonius numerous allusions to the early years of his life, and by the careful use of other more or less contemporary material we can fill in a fairly complete picture of the education of a Gallic youth in Sidonius' day.

The earliest education, the knowledge of the alphabet and of simple moral precepts, was given in the home.[3] Thence about the age of six[4] the boy would enter the school of the *grammaticus*. Here the works of the ancients were read: Homer and Menander, says Ausonius, Horace, Virgil, and Terence, and of

---

[1] Dalton, i, p. liv.

[2] The recovery of the country is best illustrated by an important comparison: the two principal symptoms of the depression of economic life in Gaul as seen by an intelligent observer in A.D. 430 were (i) the flight of the agriculturists, (ii) the increase of the Bacaudae (Salvian, *de Gub. Dei*, v. 24–6). Neither of these receives the smallest mention in Sidonius. It is notable that the word Bacauda does not occur once in his works.

[3] *C.* xxiii. 204–9; cf. Paulinus of Pella, *Eucharisticon*, 60–7.

[4] Ib. 72.

Historians, Sallust.[1] The pagan mythology was also carefully studied,[2] and this, as we shall see, has important effects upon the contemporary poetry. It is probable that the list of authors read as given by Ausonius is in substance accurate for Sidonius' own day. Menander's *Epitrepontes* was one of the works read over by Sidonius with his son Apollinaris;[3] Homer, on the other hand, who was studied by Paulinus of Pella,[4] does not seem to have been a school-book in Sidonius' day.[5] Horace and Virgil are both extensively quoted by him,[6] and of a friend he says that he had Virgil 'caned into him'.[7] In another passage he chaffs a grammarian for sitting in his gown in the blazing heat expounding Terence's *Eunuchus*,[8] and the *Hecyra* was another of the books which he studied with his son.[9] In a letter addressed to him the latter is expected to see the relevance of a comparison with Spartacus, and he is also referred to Daedalus, to Theseus, and to Proteus.[10] From this we may infer that Sallust's *Histories*[11] and the study of ancient mythology still formed part of the regular curriculum. Cicero too, though omitted in Ausonius' list, was certainly studied in the Gallic schools of this period.[12] Reminiscences and imitations of Statius are

---

[1] Ausonius, xiii (*Protrepticus*), 2. 45–65. Cf. Haarhof, pp. 56–7.

[2] Augustin., *Conf.*, i. 14. 23.   [3] *E.* iv. 12. 1.   [4] Paulin. Pell., *Euch.*, 73–4.

[5] There is one casual reference to Homer in Sidonius' prose (*E.* v. 17. 1) and two or three in his poems (*C.* ii. 185; ix. 217; xxiii. 135; *E.* ix. 15. 1, vers. 48). Sidonius cannot have been a very competent Greek scholar or he would hardly have made such glaring mistakes in the quantities of Greek names ('Euripĭdes', *C.* ix. 234; 'Fĭloctetes', ib. 156, and even 'phȳsicus' (*C.* xv. 101) and 'philosophus' (ib. 182, 187). For a friend's knowledge of Greek see *C.* xxiii, 228–32 with Stein's comments (p. 547, n. 2).

[6] See E. Geisler, *ap.* Lütjohann, pp. 353–416.

[7] *E.* v. 5. 2. On corporal punishment in Gallic schools see Ausonius, xiii (*Protrepticus*), 2. 25. Cf. *E.* ii. 10. 1 and *infra*, p. 11, n⁴.

[8] *E.* ii. 2. 2.   [9] *E.* iv. 12. 1.   [10] *E.* iii. 13. 10.

[11] Cf. 'Crispus uester', *E.* v. 3. 2.   [12] *E.* v. 5. 2. Cf. *C.* vii. 175.

frequent in Sidonius' poems, and in more than one passage he speaks of him with warm admiration:[1] it is not impossible that it was in the school-room that he made his acquaintance.

Whether Claudian, to whom Sidonius is so deeply indebted in the *Panegyrics*,[2] and Pliny the Younger, of whom there are many reminiscences in the letters, were read as school-books, we cannot say. That Sidonius does not mention them in writing of his education may be accidental; but it can hardly be an accident that neither the Bible nor any of the Christian fathers is mentioned as being studied in the school-room.[3] There is no hint in Sidonius' work that he received any religious instruction at all: the education of the lay-schools still remained a pagan education; the principles with but slight alterations of detail remained as they had been laid down by Quintilian three centuries before.

At the age of sixteen the grammatical was succeeded by the rhetorical education.[4] The tradition of rhetorical training for the youth was of great antiquity among the Romans. It arose from the fact that in early days success in public speaking was for a young and ambitious man the swiftest way to fame and high position, and such was Roman conservatism that, though the fact of free speech had long disappeared, the training originally designed for it remained. The student was given some set subject on which to speak. He might be

[1] Cf. 'Papinius noster', *C*. xxii, ep. 6, and *C*. ix. 226.

[2] Geisler has collected from the first seven poems seventy-three passages in which Sidonius is seen to have imitated Claudian.

[3] They are equally absent in the ideal education attributed by Sidonius to the emperor Anthemius, *C*. ii. 156–92; Fertig, i, p. 6.

[4] Paulin. Pell., *Euch.*, 121. Cf. *E*. iv. 21. 4, 'hic te imbuendum liberalibus disciplinis grammatici rhetorisque'.

asked to put words into the mouth of Menelaus sur-
veying the flames of Troy, or to compose a speech for
Thetis lamenting·over the dead body of Achilles: these
were 'dictiones ethicae'.[1] Or the subject might be
rather of a moral nature, such as the right of a man
to sell the burial-place of his father in order to pay
a gaming debt.[2] These were 'controuersiae'. Sometimes the subject was purely fanciful: thus one of
Libanius' *Declamations* is the lament of a parasite, who,
having hired a hippodrome horse to take him to dinner
saw his mount mistake an altar in the street for a
turning post and gallop him back home.[3] Such novel-
ties are, however, the exception. As a general rule the
subjects were traditional; it has been remarked that
'the precepts and examples which we find in Seneca
the rhetorician, are almost identcial with those of Ennodius at the end of the fifth century'.[4] And there is
another point to add to this; being traditional they bore
little or no relation to the conditions of the time. Thus
Sidonius reminds the young Burgundio that the day
is approaching when he must deliver his 'thema', and
the 'thema' is the praise of Julius Caesar.[5] It was indeed the rhetorician's boast, that he lived in the past;
one of them who is compelled to write on the affairs of
his day begs his readers to excuse his trifling.[6]

The education provided by the *rhetor* did not, however, stop at declamation. There were philosophical
classes, in which Plato and Aristotle were studied,[7] and
Sidonius' master, Eusebius, seems to have held a kind

---

[1] Ennodius, *Op.*, cdxiv and ccxx.   [2] Ib. cclxi.
[3] Libanius, *Decl.*, xxviii.   [4] Haarhof, p. 70; cf. Dill, pp. 426–7.
[5] *E.* ix. 14. 7. On declamation in Sidonius' day see also *E.* i. 4. 3.
[6] Dio Chrysostom, xxi. 11 (p. 300, ed. Dindorf).
[7] For the study of Plato see *E.* iii. 6. 2.

of 'seminar'[1] at which the pupils discussed the *Categories* of Aristotle; it is rather startling that their speculations were stimulated by flogging.[2] Sidonius also attended a class with the great Claudianus Mamertus, and years later, when Claudianus died, he wrote a consoling letter to his great-nephew, Petreius, in which there are some interesting details about contemporary philosophical studies. He tells us how groups of students would visit the great man, how they would ask his opinion of different points; how he would invite them to state one point of view and would comment upon it himself, and how he would urge the pupils to combine their propositions into syllogistic form.[3]

Under philosophy was included much that we should separate from it, for it was not only the relation to reality of the principles governing sciences that were studied under the name of philosophy but the sciences themselves; thus Sidonius in describing the advanced studies of Gallic nobles in geometry, arithmetic, and astrology, says that they are all parts of philosophy.[4] Astrology, though condemned by Sidonius in later life,[5] seems to have been very popular. Claudianus Mamertus, Leo the great jurist of Narbonne, Magnus, consul in 460, and Lampridius, Sidonius' poet friend from Bordeaux, were all keen students,[6] and the rhetorician Lupus knew well his Julianus Vertacus and his Fullonius Saturninus, the chief text-books on the subject

[1] Eusebius' pupils attended his house—'Eusebianos lares', *E.* iv. 1. 3. In some places, however, philosophical lectures were given in a hall called 'Athenaeum' (*E.* ix. 9. 13; cf. *E.* ix. 14. 2).

[2] *E.* iv. 1. 3.

[3] *E.* iv. 11. 2, 3.

[4] 'membra philosophiae', *C.* xiv, ep. 2. Cf. *C.* xxii, ep. 3, and *E.* v. 2. 1.

[5] *E.* viii. 11. 9, 'non solum culpabile sed peremptorium'.

[6] See *E.* iv. 3. 5; iv. 11. 6, ver. 9; v. 2. 1; *C.* xiv, ep. 2.

used in the Gallic schools.[1] Sidonius himself is perfectly at home in the complicated technical terminology and talks glibly of Climacterics, and the like. Such expressions as—'super diametro Mercurius asyndetus, super tetragono Saturnus retrogradus, super centro Mars apocatastaticus'—fall from his pen; use of them demands at least more than a superficial knowledge of the subject.[2]

Thus a young man who had followed the courses with attention, would go out into the world with at least a large all-round training in his own language, and a considerable, if somewhat stereotyped, knowledge of the theory of argumentation. How many-sided the education was, is well seen from a letter of Sidonius[3] written to an old schoolfellow, who seems to have been the cleverest boy of his year and to have assisted the master in teaching his less able contemporaries. From him, Sidonius says, pupils received instruction in epic, comic, and lyric poetry; in history, satire, grammar, panegyric, philosophy, epigram-writing, and law. Allowing for exaggeration, it is a broad-based humanism which is here described, in fact what Sidonius calls a 'liberalis disciplina'.[4]

Nor was the boy kept all day at his books; ball games of some sort were played, and Sidonius himself tells that he was as fond of ball as of books.[5] There

---

[1] E. viii. 11. 10; cf. C. xxii, ep. 3.

[2] E. viii. 11. 9. On astrology in Sidonius see Miremont in Rev. Ét. anc., xi (1909), pp. 201–13.

[3] E. iv. 1, addressed to Probus.

[4] E. iv. 21. 4; v. 5. 2. Compare the summary of such an education in C. xxiii. 211–12, 'quidquid rhetoricae institutionis, quidquid grammaticalis aut palaestrae est'.

[5] E. v. 17. 6; iv. 4. 1. For Roman ball games see Marquardt, Privatleben, ii, pp. 515–21 (French translation).

were also sports, running and jumping; and indoor games with the draught-board (*tesserae*).

In his vacations[1] he would be taught hunting and fishing: the emperor Avitus had been in youth a clever trainer of hawks,[2] and at the age of about fourteen had surprised his father by bringing home a wolf which. he had killed.[3] Sidonius mentions his own hunting expeditions with his friend Faustinus,[4] and probably as a boy acquired his knowledge of fishing with night lines.[5] Ecdicius, as a child, was famed for his skill with hawk and hound, horse and bow.[6]

When we know so much about the curriculum of Sidonius' education, it is rather a disappointment that we do not know at what place he received it. But in this we are introduced to a difficulty which accompanies us throughout our investigation into his life. We are, broadly speaking, dependent for our knowledge of it entirely on Sidonius himself, and, as he valued his letters rather for their style than for their material, he did not worry about withholding biographical facts from posterity. Here his silence has set us a problem to which we can only give a conjectural solution. The older writers suggested Clermont, a view to which they were led by their initial error in imagining that Sidonius was an Arvernian by birth.[7] There certainly was a school there, but it is not very likely that Sidonius was educated at it; his Arvernian connexion seems to date from a later period in his life. Lyons, Sidonius' birthplace, has been the usual choice of the biographers[8] and

---

[1] Our scanty evidence for Roman school holidays is collected by Haarhof, pp. 108–11.   [2] *C.* vii. 203–6.   [3] Ib. 177–82.   [4] *E.* iv. 4. 1.
[5] *C.* xxi. 1, 2.   [6] *E.* iii. 3. 2. On hawking see *E.* v. 5. 3.
[7] Savaron in ed. 1599, p. 7.
[8] As Chaix, i, p. 14; Hodgkin, ii, p. 299; and Dalton, i, p. xv.

it is on *a priori* grounds quite probable. Lyons was, however, rather a commercial than an academic town.[1] It is true that it had the Martyrium of Irenaeus[2] which may have been a Church school; yet in the fourth century it shared a professor with Besançon,[3] and, if it only had a part-time teacher, we may infer that the standard of education there was not very high.[4] Remembering that among his schoolfellows were Probus and Magnus Felix of Narbo, one may doubt whether they would have made the long journey to Lyons if they could have had a better education nearer home.[5] Eusebius, as we have seen, was Sidonius' philosophical tutor, and it has been shown to be highly probable that he taught, not at Lyons but at Arles.[6] There is another argument which makes it very probable that it was at Arles that Sidonius received at least his rhetorical education. A well-known passage describes how as a youth he stood beside his father, then praetorian prefect, and witnessed the ceremonies performed in honour of the new consul Astyrius.[7] There can be no doubt that this scene occurred at Arles, which had been since 408 the residence of the Gallic praefects. If at the age of seventeen Sidonius was at Arles at a time when his father was living there as prefect, it is fair to conclude that the latter part of his education was acquired there. Whether he received both his grammatical and his rhetorical education at the same place is another ques-

[1] Allard, p. 7.

[2] Boissieu, *Inscriptions de Lyon*, p. 548, quoted by Haarhof, p. 46. Cf. 'caterua scholasticorum', *E.* v. 17. 6.

[3] Ausonius, *Grat. Actio*, viii. 7. 31; Jullian, *Ausone et Bordeaux*, p. 52.

[4] The inscription mentioned above is very illiterate.

[5] *E.* iv. 1. 1; *C.* ix. 330; Germain, p. 5. Besides these two we know the names of Avitus (*E.* iii. 1. 1) and Aquilinus (*E.* v. 9. 3).

[6] By Coville, p. 40, quoting *Vita Hilarii*, xi.   [7] *E.* viii. 6. 5.

tion that we cannot certainly decide. Both Allard[1] and Coville,[2] the most recent students of his life, are inclined to believe that he learnt grammar at Lyons and rhetoric at Arles. In favour of this is the fact that one of his schoolfellows, whose name he records to us, was also from Lyons: on the other hand, among the subjects in which the brilliant Probus of Narbo assisted his contemporaries are mentioned distinctly grammar and the study of poets.[3] Certainty is hardly possible to reach, but from the latter passage one would be inclined to conclude that the 'grammatical' education under Hoënus[4] was also accomplished at Arles. Whether Claudianus Mamertus taught at Arles, whether indeed he was a recognized professor at all, is quite uncertain. When we next hear of him he is residing at Vienne.[5] Perhaps Sidonius on his way home from Lyons paid him a visit and there received informal philosophic instruction.

If some space has been taken up with a fairly detailed account of Sidonius' education, it must not be considered as wasted. It is obvious that the system under which he was educated must have had the most vital influence upon his manner of thought and his expression of it in verse and prose; and we can make no approach to the literary criticism of his work, unless we realize what he was trying to do. Sidonius was not a leader of revolt in the world of thought; he was a conventional man of the times, very much so. He did not resist the critical standards of his day, he conformed to them; and we shall not go far towards understanding what were the conventional standards to which he conformed, unless we appreciate the education which

---

[1] p. 7.   [2] p. 40.   [3] Aquilinus, see *E.* v. 9. 3.   [4] *C.* ix. 313.
[5] *E.* iv. 11. 6, vv. 19–21.

moulded them. And if we can deduce some of these standards, we may begin to understand not only the mind of Sidonius, but that of the age in which he lived, the fifth-century mind. For Sidonius has been willingly thrust forward by his contemporaries to be their wit-ness. Claudianus Mamertus is proud to learn that Sidonius has deigned to accept the dedication of the *De Statu Animae*:[1] Leo, the jurist of Narbonne, told him that he was made for better things than letter-writing:[2] Tonantius Ferreolus thought him comparable to the greatest of poets,[3] and these men were not fools.

Hard words have often been said about this fifth-century education, and the practice of rhetorical declama-tion has been especially singled out for censure. Yet it may well be defended; a knowledge of how to stand upon one's feet and argue is always useful, and the de-tractors of rhetoric have forgotten that there are other places for it than the republican rostra. Men who could argue and arrange their ideas in logical form must always have been good instruments for pushing business through at councils of the municipality, of the vicar or prefect, or at the Provincial Assembly. The single public oration of Sidonius preserved to us,[4] that in which he announces the election of a certain man as bishop of Bourges,[5] is certainly modelled on the lines of the rhetorical tradi-tion, but it is not a bad speech, it has cohesion and logical arrangement, and presses forward to the con-

---

[1] *E.* iv. 2. 2. Cf. Claud. Mam.; *de Statu Animae*, Praef. and i. 1, and Germain, p. 112.

[2] *E.* iv. 22. 1, 'idoneum quippe pronuntias ad opera maiora, quem mediocria putas deserere debere'.

[3] *E.* ix. 13. 1, '. . . ut poetarum me quibusque lectissimis comparandum putes, certe compluribus anteponendum'.

[4] Apart, of course, from the verse panegyrics, which are of a somewhat different type.   [5] *E.* vii. 9. 5–25.

clusion without repetition or vagueness. To blame the declamation unreservedly and to assert its evil effects upon literature is to miss a very important point. The declamation did not perish with the Roman Empire; it survived as an instrument of education for centuries after the Renaissance; and the fact that there are school debating societies in which it is forbidden to discuss politics or religion shows that the practice of debating on subjects removed from actuality is still recognized as a part of education.

What has been observed about the survival of declamation leads one on to consider a further point. The scholar who sets out to criticize the principles of fifth-century education is in some danger of having his complacency pricked by the thought that he has met these principles elsewhere. An education which begins with commentary upon classical poets and orators, and ends with the discussion of Plato and Aristotle is surely familiar enough to us in the twentieth century.

Fortunately, however, we are spared the necessity of deciding whether the Gallic curriculum of the fifth century was good or the Oxford classical tradition of the twentieth bad. There is after all a fundamental difference between the two principles; and it is just this difference which illuminates the history of fifth-century thought. The modern humanistic education investigates the debt of a present civilization to a predecessor founded upon different principles; it traces the development of this civilization, its reaction to its environment, and the influence which the environment had upon its greatest men. It applies the lessons so learnt to conditions based on its own environment, it endeavours to deduce first principles of human activity, whether in

ethics, politics, or aesthetics. In its essence it is opposed
to a practical training. But in this lies the great difference between it and the education of the fifth century;
that education was itself practical, its object was actually to train men to speak like the ancients, to write like
the ancients, and to think like the ancients. In working
along the lines of the classical tradition it lacked the
advantage of using as a medium a language recognized
as dead. Men were taught to think and write like
Cicero and Virgil, and though, as we shall see, interest
in literature was declining, the age can boast of an
unexpectedly large output of literary work.

The aristocratic class consisted of men with time
hanging heavily on their hands, and they amused themselves with literature. South Gaul was a 'nest of singing
birds'. With the exception of Sidonius' own work not a
vestige of this abundant literary effort is handed down
to us, and how far he succeeded in imitating his models
we know well. How successful the other imitators
were, we do not know; but the point is not important.
Their works were criticized on the assumption that
they were imitations; that is the important point. Instances of this are abundant in Sidonius. Very instructive is the praise lavished on Consentius, a poet friend
from Narbo. He is said to surpass practically every
poet and prose writer down to the silver age, but no
further.[1] Leo and Polemius are both declared to surpass Tacitus.[2] Proculus, a Ligurian poet, is said to

[1] C. xxiii. 97-169. It is significant that Sidonius only once mentions his
immediate predecessors, and then when he is declaring that his poetry is inferior
to any poet who ever lived including them (C. ix. 274-301). There is a single
mention of Ausonius (E. iv. 14. 2), none of Claudian (outside the passage in
C. ix).

[2] E. iv. 22. 2; iv. 14. 2. Cf. C. xxiii. 446-54; E. ix. 15. 1, vv. 20-34 (Leo
as a poet).

challenge Virgil.[1] Sidonius himself apologizes to a correspondent for not writing 'more ueteris studii'; his work will now be, he says, 'mere rambling, pointless stuff'. What his correspondent wants is something antique.[2] Lampridius' work is praised for its imitation of Pindar and Horace;[3] the last instance is particularly valuable, for Lampridius was dead when this was written, so all suspicion of flattery is absent. Finally we may quote the words addressed to Sidonius by Claudianus Mamertus. 'Eruditissime uirorum',[4] he says, 'ueteris reparator eloquentiae.' Almost every word here is a commentary upon the literary history of the fifth century. *Eruditio—uetustas—eloquentia*—here are its ideals put before us in the clearest light from the mouth of one of its most acute intellects.

In truth, during the last centuries of the Roman Empire, there arose what one can only call a cult of the antique. It is seen throughout the works of the later Greek sophists, and it is no accident that to St. Augustine we are indebted for much of our information about the early Roman Republic.[5] That the contemporary chronicles were careful in almost all cases to go back to Adam, or at least Romulus, is an historical fact no less important than the events of their own time for which they are quoted by modern historians. It is a symptom of the universal antiquarianism, of the looking back towards the past. This cult of the antique is even seen in the laws promulgated to the Roman world. Thus Constantine declares: 'Venientium

---

[1] *E.* ix. 15. 1, vv. 44–9.   [2] *E.* viii. 16. 2, 'dictio uetuscula'.

[3] *E.* viii. 11. 7.

[4] *de Statu Animae*, Praef. Cf. *E.* ii. 10. 5; Ruricius, *Ep.* ii. 26. 8, 'eloquentiae flore'.

[5] So too Macrobius, *Sat.* iii. 14. 2, 'Vetustas quidem semper, si sapimus, adoranda est'.

est temporum in disciplina stare ueteribus institutis.'[1] According to a law of Arcadius and Honorius—'Mos retinendus est fidelissime uetustatis.'[2] And a contemporary of Sidonius, the emperor Majorian, extols the 'uetus prouidentia dispositioque maiorum quam in omnibus sequimur atque reparamus'.[3]

This imitative education had a further consequence: it was not easy to imitate Virgil's ideas, they were the ideas of his own day, subjected to the conditions of his own day; it was not easy under the rule of a barbarian Master of Soldiers to imitate the ideas of Cicero living under a free republic. But where it was difficult to imitate the matter of antiquity, it was not so hard to imitate the manner. The kindest criticism of fifth-century educational principles would be that they set more store on the training of the intellect than on the intellect itself. It is no less true to say that they taught men to think and write and gave them nothing to think or write about. They exalted form at the expense of matter. When Sidonius says that one of his fellows writes poetry better than Virgil, it is by no means certain that he is not speaking the truth. Let us assume that hiatus and unfinished lines are regarded as blemishes in hexameter writing, then a writer who avoids them is *pro tanto* better than Virgil.[4] It is quite likely that from this point of view some authors did surpass their predecessors. An example is to hand; from a study of the classical Greek orators a rule can be inferred, that they followed certain accentual laws in the fall of the sen-

---

[1] *Cod. Theod.*, v. 20. 1.   [2] Ib. iv. 4. 4.

[3] *Novell. Maioriani*, ii. 2; cf. Seeck, *Untergang*, i, p. 442.

[4] According to Baret, p. 101, 'On remarque moins de licences, en fait de métrique et de prosodie, dans les vers de Sidoine, que dans les poésies de Vergile.'

tence; but they often broke the rule, and perfect sinlessness is not attained till the sixth-century historian Agathias.[1] That is why it is not inconsistency in Sidonius to say on the one hand that so-and-so surpasses Virgil, and on the other that the works of the ancients were privileged and beyond criticism.[2] In one way they were, in another not.

The exaltation of form at the expense of matter is very clearly seen in the writings of Sidonius himself. It is well seen in the letter to Claudianus Mamertus in which he comments upon the *De Statu Animae* and a hymn which Claudianus had sent to him. The letter is in the nature of a review or appreciation, and from our point of view it must be acknowledged to be about the worst review that was ever written. It abounds in the shallowest of comparisons between Mamertus and various thinkers of the past.[3] It praises this philosophical work because the caesurae are properly arranged[4] and because Mamertus has reintroduced old words from classical authors. The criticism of the hymn is even more instructive. Its peculiar merit, says Sidonius, is that it observes each foot in the metre, each syllable in the foot, and each emphasis in the syllable. 'It seems mere play to you', he says, 'with your tiny trochees and tinier pyrrhics to surpass in effect not merely the Molossian and anapaestic ternary, but even the quaternary, the epitrite and Paeonian rhythms.'[5] A work of Faustus of Riez is noticed as being 'bipertitum sub dialogi schemate, sub causarum themate quadri-

---

[1] See C. Litzica, *Das Meyersche Satzschlussgesetz*, 1898.

[2] *E.* iv. 3. 1; cf. iii. 8. 1.

[3] There is plenty of such stuff as—'simulat ut Crassus, dissimulat ut Caesar, suadet ut Cato, dissuadet ut Appius, persuadet ut Tullius'.

[4] *E.* iv. 3. 3.   [5] Ib. § 8.

pertitum'.[1] The important point is not so much that the criticism came from Sidonius as that the critic was seriously admired and approved as a man of intelligence by Claudianus and Faustus, the two men of the time who really did have something to say.

We see then that the education of the fifth century was an antiquarian education, and as such primarily a pagan education. This goes far to explain not only the literary style of Sidonius himself, but the swift decay of culture that followed him. Culturally the Gallic society was living precariously upon its past. It was only in the Church that education encouraged and stimulated original thought; Faustus and Claudianus were both products of ecclesiastical education as were in the next generation Ruricius and Avitus of Vienne. The Church schools survived the shocks of the fifth and sixth centuries, the lay schools did not; and this is a fact which has the very widest consequences in the history of culture. Faustus and Claudianus are the forerunners of Scotus and the schoolmen, but Sidonius left no successor.[2] From one point of view it would not be wrong to call Sidonius the last of the Romans in Gaul, for with him the Roman tradition was broken.

[1] *E.* ix. 9. 10. Sidonius' knowledge of metrical technique is well seen in *E.* ix. 13. 2, vv. 1–13, where a description of various Horatian metres is strangely squeezed into verse. On his interest in *versus recurrentes* and the like see *E.* ix. 14. 4–6.

[2] This point is made by Stein, p. 546.

II

SIDONIUS AND AVITUS

SIDONIUS does not tell us at what age he left school, and we cannot appeal to any certain indication of a school-leaving age at that time. The educational law of Valentinian forbids a student to remain at the schools of Rome beyond his twentieth year,[1] and it is probable that in the provinces the leaving age was lower.[2]

Not later than 452, then, Sidonius returned to his family at Lyons; and some time afterwards it seems that he published a volume of poems. In his old age Sidonius regretted that they had ever seen the light. He was young at the time, and they were written in the first heat of youth. Possibly they were scandalous. At any rate, Sidonius hoped that 'the greater part of them might be buried in silence'. It is hard to know whether to be pleased or sorry that he has his wish.[3]

About the time of his first literary venture, another still more important event occurred in Sidonius' life—he married.[4] His bride was Papianilla,[5] daughter of Eparchius Avitus[6] an Arvernian of moderate means but of distinguished family.[7] She seems to have been a

---

[1] *Cod. Theod.*, xiv. 9. 1.

[2] See Ausonius, xvi (*Professores*), 18. 10–11, 'tum pueros . . . formasti rhetor metam prope puberis aeui'.

[3] *E.* ix. 16. 3, vv. 41–4; cf. *E.* v. 21.

[4] Tillemont (*Mém. Eccl.*, xvi, p. 748) first pointed out that Sidonius was already married when Avitus became emperor (see *C.* xxiii. 430 and *E.* i. 3. 1). Moreover, *C.* xvii. 3, which, as will be shown (p. 144, n. 1) can hardly have been written later than 469, proves that the marriage took place not later than Oct. 452.　　　　　[5] *E.* v. 16; Greg. Tur., *Hist. Franc.*, ii. 15 (21).

[6] For the name see de Rossi, i, 795.

[7] Avitus' family—*C.* vii. 155–7; Greg. Tur., *Hist. Franc.*, ii. 10 (11). His poverty—*C.* vii. 568, 'pauper legeris'.

distant relative of Sidonius on his mother's side.[1] From her he obtained as dowry the Arvernian property called Avitacum, which henceforward became a second home to him, more pleasant, as he tells us, than his own ancestral estate at Lyons.[2] This marriage had decisive consequences for Sidonius; by it he became son-in-law to one of the most prominent imperial subjects in Gaul. With such a patron at his back he might confidently enter on the political career in which his family had for the last three generations won such brilliant prizes. With his marriage to Papianilla, Sidonius becomes an actor on the stage of world history; his biography can no longer be kept separate from the general history of the Roman Empire. At this point, therefore, we are justified in making a survey of the political state of the West in the years 453–5.

Valentinian III had sat since 425 on the imperial throne: he had succeeded in infancy, and now in 453 he was a feeble degenerate of thirty-four; insane and scarce a man, Sidonius called him.[3] The real direction of affairs was in the hands of Flavius Aëtius, Patrician and Master of the Soldiers. To Gaul, at least, he had not failed in his duty: he had checked the attempts made by the Visigoths to push forward the frontiers

---

[1] Writing to another Avitus, presumably a connexion of the emperor, Sidonius says—'matribus nostris summa sanguinis iuncti necessitudo', E. iii. 1. 1. Coville's statement (p. 41) that Papianilla was the first cousin of Sidonius is unsupported by any evidence.

[2] E. ii. 2. 3, 'Auitacum . . . nomen hoc praedio, quod, quia uxorium, patrio mihi dulcius'. The inference from this passage seems indisputable, yet it is curious that what was presumably from its name the family property went to a daughter as dowry, though there were two sons, Ecdicius and Agricola. The former is mentioned as a visitor to Avitacum (E. ii. 2. 15) and he had a town house at Clermont (E. iii. 3. 5), but where his estate was we cannot say. The other son, Agricola, had a property near Lyons (E. ii. 12. 1).

[3] 'Semiuir amens', C. vii. 359. It is therefore not a little curious that in another passage Sidonius calls him 'pius princeps', C. xxiii. 214.

assigned to them by the treaty of 418, he had settled the Burgundians as federate barbarians in Savoy, and in alliance with the Visigoth he had defeated Attila on the Mauriac plains (451). His detractors had even accused him, as it seems, of abandoning Italy to save Gaul in 452.[1] Under his rule the Gauls, though weighed down with tribute, had remained steadily loyal.[2] There may well have arisen complaints about the incapacity of their emperor, and men doubtless hinted that, with the Theodosian dynasty in such obvious decay, it might be better to found a new imperial house from the virile stock of the Gallic nobility,[3] yet it is in the highest degree significant that no attempt was made to imitate the events of 410. Jovinus and Sebastian had no successors: throughout the long reign of Valentinian no Gallic pretender arose.[4] The Theodosian dynasty still ruled unshaken in the person of its last male representative.

The Visigoths, who had been imperial *Foederati* for nearly forty years, had been making from time to time various more or less successful attempts to enlarge their territory at the expense of the empire. In 451, however, in spite of some hesitation, they had supported Aëtius in the struggle against Attila. Their king, Theodoric I, had fallen in the battle, and had been succeeded by his son Thorismund, but in 453 the

[1] This seems the most plausible explanation of the obscure passage in Prosper Tiro, 1367 (i, p. 482).

[2] *C.* v. 446, 'Gallia continuis quamquam sit lassa tributis.' It is noticeable that Sidonius usually speaks well of Aëtius, even when praising his subordinates Avitus and Majorian (*C.* v. 275–93; vii. 231).

[3] The evidence for this feeling is derived from the necessarily dark hints dropped by Sidonius in his panegyrics. See *C.* v. 356–63; vii. 537–47.

[4] It is noticeable that Sidonius had hard words for all the fifth-century usurpers (*E.* v. 9. 1) even though one of them (Jovinus) was the father-in-law of a friend (*C.* xxiii. 173).

latter had fallen victim to a domestic conspiracy. He
was murdered by his brothers, the eldest of whom,
Theodoric II, succeeded him. The contemporary
chronicler, Prosper Tiro, who at Marseilles was close to
the events which he describes, relates as motives for the
murder that Thorismund 'wished to disturb the peace
between the empire and the Goths',[1] and this is very
probably correct. It might then have been thought
that the new king, a lover of Roman literature and
Roman ways,[2] would act up to the sentiments of his
great-uncle Athaulf, and be 'the restorer, not the sub-
verter, of the Roman name'.[3] Thus for the time some
sort of *status quo* seemed possible. But if Aëtius died?
Or Valentinian?

We can hardly doubt that such questions were in the
minds of many of the Gallic nobles—and among them
Eparchius Avitus. His career had been distinguished,
and he had already been praetorian prefect of Gaul
(perhaps 439–40).[4] On several occasions he had been
sent on important missions to the Visigoths, where he
was a guest friend of the royal house. He had been one
of Aëtius' most successful generals in the Gallic wars of
430–45. A man of some fifty years of age now living
in retirement on his Arvernian estate,[5] what was his
future to be?

Theodoric too may well have speculated as to his
own future, if there were any changes in the *status quo*.
He may well have thought that if there was a Gallic
emperor he might use Avitus as a second Attalus.
For Avitus was a bold warrior, but not a man of

---

[1] Prosper Tiro, 1371 (i, p. 483).   [2] *C.* vii. 495–9.
[3] Orosius, vii. 43. 4–6.
[4] Sundwall, p. 23.
[5] Cantarelli, p. 56; *C.* vii. 378–81.

very acute intelligence,[1] in fact an excellent figure-
head for an emperor, behind whom Theodoric might
hope to play with more success the role attempted
forty years before by Athaulf.

It was at about this time that Sidonius made a journey to the Visigothic capital of Toulouse,[2] in order to
make certain requests of Theodoric.[3] A personal interview between a member of his family and the Visigothic king was of course a matter of interest to Avitus.
He had himself long ago taught Theodoric the laws of
Rome and the poems of Virgil,[4] but it was important
to know what his character was and what his attitude
might be towards the Gallic aristocracy now that he
was king. Agricola, son of Avitus and brother-in-law
of Sidonius, wrote frequent letters to Toulouse demanding information.[5] Sidonius complied with the request,
and his letter (*E.* i. 2), which is the earliest, is also one of
the most interesting of the collection. If it did not perhaps tell Agricola all that he wished to know, it told
him much that was important. The new king was an
affable, kindly man, so much so, says Sidonius, as to be
exempt from the unpopularity that usually attends a
king.[6] Sidonius commented upon his agreeable banquets 'where you could see the elegance of Greece, the
abundance of Gaul, and the celerity of Italy'.[7] He
added a detail likely to be of great interest to the recipients: the new king's Arianism, always a stumbling-
block to the Gallo-Roman provincials,[8] was far from

---

[1] 'Homo totius simplicitatis', Victor Tonnennensis, ii, p. 186.
[2] *C.* vii. 436.   [3] *E.* i. 2. 8.
[4] *C.* vii. 495–9.
[5] 'Saepenumero postulauisti', ib. § 1.
[6] *E.* i. 2. 1; cf. ib. § 8, 'dico quod sentio timet timeri'.   [7] Ib. § 6.
[8] Cf. Woodward, *Christianity and Nationalism in the later Roman Empire*, p. 79.

enthusiastic and he suspected that it was dictated more by habit than by conviction.[1] The new king evidently made a good impression on Sidonius,[2] rather too good for some modern commentators, who have not failed to observe that 'ciuilitas' is a curious word to use of a fratricide.[3] It almost looks as though a certain moral toleration was extended to barbarian kings in their domestic affairs.

Soon after Sidonius' journey two events occurred which completely subverted the *status quo*. On September 4th, 454,[4] Aëtius was murdered by Valentinian, and six months later[5] Valentinian himself was assassinated by two of Aëtius' comrades;[6] they had been instigated, as it appears, by Petronius Maximus,[7] a Roman senator who had already been twice consul and twice praetorian prefect, the two highest posts that a subject could hold.[8] What effect the assassination of Aëtius alone would have had upon the barbarians of the West, we can hardly say; the swiftly succeeding deaths of the great general and the last male heir of

[1] *E.* i. 2. 4.

[2] It is very unfortunate that we cannot date this letter at all precisely in its relation to the political situation of the time. It cannot be earlier than the end of 453, when Theodoric succeeded (cf. *E.* vii. 12. 3 with *Chron. Gall. a. dxi.*, 621, i, p. 663), and it is obviously quite early in his reign. But whether it is before or after Aetius' murder, it is quite impossible to say. I have made in the text the confessedly arbitrary supposition that it precedes the murder. But it is just possible that Sidonius accompanied Avitus to Toulouse and wrote the letter from there in 455.   [3] As Dalton, i, p. cxliv.

[4] *Addit. ad Prosper. Haun.* (date), 570 (i, p. 303).

[5] *Cont. Prosper.* (March 16th, 455), 27 (i, p. 490).

[6] 'Bucellarii'; *Addit. ad Prosper. Haun.*, 572 (i, p. 303). Whether it is more than a coincidence that both the murderers were Goths one cannot say (see Mommsen in *Hermes* (1901), p. 540).

[7] John Ant., fr. 201, 4 (*F.H.G.*, iv, p. 615); Jord., *Get.*, xlv. 235. For the sequence of events between the death of Aëtius and the accession of Avitus see Cessi in *Archivio della Reale Società romana di storia patria*, x (1917), pp. 161–204, and Stein, p. 519, n. 2.

[8] *E.* ii. 13. 1; *C.I.L.*, vi. 1749 (=Dessau, 809); Sundwall, pp. 104–5.

the house of Theodosius produced considerable distur-
bance, which must have been intensified by the news
that Petronius Maximus had actually succeeded to the
throne. The independent barbarians, freed from the
strong hand of Aëtius, renewed their attacks, and
the federates, who seem to have regarded their treaties
as binding them to the Theodosian house rather than
to the empire, everywhere renounced their allegiance.[1]
The Saxons attacked the coasts of north Gaul, the
Alamanni crossed the Rhine, the Franks pushed forward into Germania Prima and Belgica Secunda,[2] the
Gepidae and Burgundians began to move;[3] in Spain
the Suevi ravaged Cartagena.[4] The two kings, however, whose attitude was the empire's chief concern,
were, of course, Theodoric the Visigoth and Geiseric
the Vandal. It is easy to conceive the annoyance with
which Theodoric must have viewed the election of the
Italian emperor, and it is likely enough that his annoyance was shared by the Gallo-Roman aristocracy and
Avitus himself. In their fury it seems that the Goths
had actually formed the project of marching on Rome.[5]

No less threatening was the attitude of Geiseric:
immediately after the news of Valentinian's murder he
had acted as though the *foedus* no longer existed, and
had taken possession of those parts of Africa which had
hitherto belonged to the empire. With the aid of his
powerful fleet he reduced Sicily, Corsica, Sardinia, and
the Balearic Islands,[6] and prepared to attack Rome itself.

Such dangers abroad might have daunted the
strongest heart, but there was worse as home. After

---

[1] Cantarelli, p. 49.   [2] C. vii. 369–75.   [3] See p. 26, n. 8.
[4] Hydatius, 168 (ii, p. 28).
[5] C. vii. 361–2, '. . . necnon sibi capta uideri / Roma Getis, tellusque suo
cessura furori'.   [6] Victor Vitensis, i. 14. 3.

the murder of Aëtius the command of his troops would naturally have fallen to his two subordinates on Italian soil, Flavius Ricimer and Iulius Valerius Maiorianus.[1] Of these Ricimer was a barbarian, the son of a Sueve and a Visigoth;[2] Majorian was a Roman from Pannonia, who, after the death of Aëtius, had been made *comes domesticorum* by Valentinian.[3] When we consider the history of the next few years, it is hard to resist the conclusion that Ricimer had already in mind the creation of Majorian as emperor with himself as Master of the Soldiers in place of Aëtius.[4] On the death of Valentinian, indeed, an attempt had been made to have Majorian created Augustus, but Maximus had frustrated this by immense bribes.[5] The consequence of these events is obvious: like Didius Iulianus two centuries before, Maximus had to face hostility abroad with an army of uncertain temper at home.[6]

Against the terror from the west Maximus acted promptly and with some ingenuity. Avitus was conciliated by being appointed to the office of *magister utriusque militiae per Gallias*:[7] this measure had the effect of taking the wind from the sails of Avitus' supporter Theodoric.

The results of this appointment, if we may believe Sidonius, were immediately seen. Avitus quickly repelled the Gepidae and Burgundians;[8] his appearance

---

[1] For their full names see Dessau, 1294 and 810; *C.* iv. *inscr.*

[2] *C.* ii. 361–2.   [3] *C.* v. 306–8.

[4] *C.* v. 266 *et seq.*, as Cessi says (p. 183), seems to imply as much.

[5] John Ant., fr. 201, 6 (*F.H.G.*, iv, p. 615).

[6] Cf. *E.* ii. 13. 5, 'ipsam aulam turbulentissime rexit inter tumultus militum popularium foederatorum'.

[7] *C.* vii. 376–8, so Sundwall, pp. 55, 60. But the appointment may have been to the office of *magister militum praesentalis*. See Ensslin in *Klio*, xxiv (1931), pp. 486–9.

[8] *Cont. Prosper.*, 5 (i, p. 304), 'At Gippidos Burgundiones intra Galliam

in Upper Germany forced the Franks and Alamanni to retire, and the Saxon incursions ceased.[1] Meanwhile his general, Messianus, had been holding the Goths in check,[2] and after settling matters on the eastern frontier, Avitus himself journeyed to Bordeaux and there made a personal appeal to Theodoric to maintain the *foedus*, as his father had done.[3] In this he was successful, and his success was a victory for the statesmanship of Maximus.

Yet it was a fruitless victory. For in May Geiseric landed at the mouth of the Tiber and prepared to attack Rome.[4] It is evident that his arrival occurred before any one had expected him. The army (or such of it as was in Rome) was undisciplined and disaffected: a panic ensued in which every one sought to fly from the city, and in the commotion Maximus lost his life[5] (May 31st, 455). No attempt was made to resist Geiseric's advance: he entered Rome, and for fourteen days the unfortunate city was subjected to systematic pillage.[6]

The course of events in Italy must have been quite to Theodoric's liking. The general demoralization had spread to the army and even to its leaders. Neither Ricimer nor Majorian had any plan or took any action.[7]

---

diffusi repelluntur.' The passage is corrupt, but 'Gippidae et B.' is an easy correction. These Gepidae, who are never heard of elsewhere, must have been part of Attila's army cut off from their fellow tribesmen in Dacia (Jord., *Get.*, xxxviii. 199).

[1] *C.* vii. 388–91. An anecdote in ps.-Fredegar (iii. 7), worthless in itself, seems to preserve a tradition that Avitus was in Trier at this time.

[2] *C.* vii. 426–7, &c.

[3] *C.* vii. 469–70, 'foedera prisca precor, quae nunc meus ille teneret, iussissem si forte, senex'.　　　　[4] John Ant., fr. 201, 6 (*F.H.G.*, iv, pp. 615–16).

[5] Cf. *C.* vii. 442–3; *E.* ii. 13. 4; Seeck, *Untergang*, vi, pp. 473–4.

[6] Prosper Tiro, 1375 (i, p. 484), 'secura et libera scrutatione'.

[7] Cf. *C.* vii. 513.

It was a golden opportunity for Theodoric to seize that position of arbitrator of the destinies of the empire, which had been for so long held by the commander of the Italian *foederati;* and the humiliation of Italy was no less an opportunity for the nobility of Gaul, who for years had endured the rule of a feeble Italian dynasty supported by the commander of the Italian federate troops with a council of state now almost entirely composed of resident Italian members.[1]

A sop had indeed been thrown to the Gallic nobility by the establishment of a kind of Gallic senate at Arles,[2] and it had been for some time the custom to appoint a Gaul as praetorian prefect[3]; but this was poor consolation for the subordination to Italy. Now, however, all was changed: the Italian dynasty was extinct, the army disorganized, and the senate scattered after the sack of Rome.[4] Now was the time for Gaul to take the lead and save the world lest the world perish.[5]

The news of Maximus' death reached Theodoric just at the time when Avitus was at his court arranging for the maintenance of the *foedus.* Theodoric at once urged him to assume the purple, and it may be that Avitus, in spite of the reluctance of which Sidonius makes so much, was by no means unwilling.[6] Accompanied by Theodoric and his brothers, Avitus came to Ugernum, the modern Beaucaire,[7] where an assembly of Gallic notables confirmed in the name of the provincials the request of the Gothic king (July 7th).

---

[1] See Lécrivain, pp. 65–7.   [2] See Carette, pp. 242–51 and pp. 450–63.
[3] Sundwall, p. 8.
[4] *C.* vii. 450–1, 'exilium patrum, plebis mala, principe caeso / captiuum imperium ad Geticas rumor tulit aures'.
[5] *C.* vii. 518–19, 'tibi pareat orbis, / ne pereat'.
[6] *C.* vii. 519–20, 579–84; Greg. Tur., *Hist. Franc.,* ii. 11 (12), 'ambisset imperium'.   [7] Marius of Aventicum, ii, p. 232; *C.* vii. 525–75.

Avitus assented, and two days later he was formally crowned with a military torque and raised to the empire by the Gallic army at Arles.[1] Theodoric supplied him with a body-guard of Goths, and he set out for Rome.[2]

With him went his young son-in-law, and other Gallic nobles, eager as he to gain official distinction.[3] That Sidonius at this time entered upon a civil service career cannot be doubted. Not long after this he refers to his 'recens commilitium' and the journey taken abroad on it. At this time when Roman citizens did not as a general rule undertake commands in the field, the word 'militare' is used in the technical sense of being a member of the imperial civil service: even the officials in the imperial secretariat ('scrinia sacra') were said 'militare in sacro palatio'.[4] One may presume that Sidonius started at the bottom of the ladder, and thus he would have accompanied Avitus as 'tribunus et notarius.'[5] Sidonius was full of hope and optimism as he entered on the new career. To a friend he wrote in the most cheerful language: 'You can,' he says, 'charge me with intrigue and have me removed from the senate, but why should not I be ambitious for a career like my ancestors? My father, father-in-law, and grandfather have all held the highest dignities, both civil and military.'[6] It was in this

---

[1] C. vii. 576–9; Hydatius, 163 (ii, p. 27); Fast. Vind. Priores, 575 and Cont. Prosper., 6 (i, p. 304). For the date see Seeck, Untergang, vi, p. 476. The accounts of Sidonius and the chroniclers cannot certainly be reconciled; but Seeck's reconstruction (in P.-W., ii, p. 2396) is plausible.

[2] John Ant., fr. 202 (F.H.G., iv, p. 616).

[3] C. xxiii. 428–31.   [4] Cod. Theod., vi. 26. 5.

[5] So Mommsen (Rede und Aufsätze, p. 134). For an example of such an office assigned to a young man fresh from school see C. xxiii. 215–16.

[6] E. i. 3. 1. This letter and the next (E. i. 4) which goes with it, are usually

hopeful spirit, far removed from the disillusionment of his later life, that he left Gaul with his father-in-law and master for the new adventure.

On his journey to Rome Avitus turned aside to Pannonia where he received the submission of the mixed Scirian and Rugian barbarians who dwelt there,[1] and it was not till September 455 that he entered Italy. Soon after this he was recognized as Emperor at Rome, and on January 1st he assumed the consulship. It had been the practice under the republic for the consuls on the occasion of their assuming office to make an official supplication to the gods for the welfare of the state;[2] under the empire prayers were also offered for the safety of the emperor. Such public prayers tended naturally to take the form of an official laudation of the emperor,[3] and in this the influence of the Greek set speech is probably to be seen.[4] When the consul was not himself a practised speaker, or when the emperor himself was consul, the laudation might be entrusted to a notable orator.[5] Such laudatory speeches were not confined to the first day of the emperor's consulship, but might be delivered on his birthday, his *vicennalia*, or other such occasion.[6] The early panegyrical orations were in prose. But in the fourth century they began to change in character. It had been the habit of panegyrists to apostrophize the goddess Rome in their addresses to the emperor,[7] but with the trans-

assigned to a date early in the reign of Anthemius, but from the tone of them I should prefer to date them to 455 (so Tillemont, *Mém. Ecclés.*, xvi, p. 199).

[1] See Alföldi, *Der Untergang der Römer Herrschaft in Pannonien*, ii, p. 100.
[2] See Mommsen, *Staatsrecht*, iii, p. 154 (French translation).
[3] Ib., vi, p. 77.   [4] As Aelius Aristides and Dio Chrysostom, *E.* ii. 1.
[5] Cf. Claudian, *Pan. Olybrio et Probino conss.*
[6] As the *Pan. Lat.* in Baehrens' collection.
[7] See Ael. Arist., Εἰς Ῥώμην; *Pan. Lat.*, iv. 13; x. 13; xi. 12, &c.

formation of the empire into a pure despotism, the
divinity of the emperor was more clearly expressed and
his separation from the rest of his subjects more sharply
defined.

Thus, Diocletian and Maximian are 'Jovius' and
'Herculius' and Christian princes in their laws habitually speak of their 'sacrum numen':[1] the emperor to
whom a panegyric was addressed had become practically
an equal of the goddess Rome. As a consequence the
panegyric became as much a hymn as a speech, and for
that reason was frequently cast into verse. Fragments
of such a verse panegyric to Diocletian and Maximian
were discovered in Egypt not long ago.[2] Even after the
emperors had become Christians, the pagan tradition
persists: the emperors are not, it is true, actually addressed as gods, though the language is not far distant,[3]
and the figure of the deified Rome remains. Rome is
personified, clothed in armour, rags, or what not,[4] and
takes part in a council of gods or appeals to another
city of the empire. This personification of the cities is
a very noticeable feature of the panegyrics of Claudian
and Sidonius: it is interesting to speculate whether it
owes anything to the allegorical sculptured figures of
great cities of the empire, which seem to have been
common about this time.[5]

The duty of delivering the panegyric to Avitus was

[1] Examples from the codes are numerous. *Cod. Theod.*, vi. 5. 2. is one of the
most curious. Compare also the military oath in Vegetius, ii. 5.

[2] See R. Reitzenstein, *Zwei religionsgeschichtliche Fragen*, pp. 46–52.

[3] Cf. Claudian, *Pan. de Tert. Cons. Honorii*, 121–2 ('gaudent Italiae . . .
oppida . . . adventu sacrata tuo'); *de Quart. Cons. Honorii*, 132–53, 602–10, &c.

[4] Cf. Symmachus, x. 3; Claudian, *de Bello Gildonico*, i. 17–127, &c.

[5] Compare the four statuettes representing the *Tychae* of Rome, Constantinople, Antioch, and Alexandria found at Rome in 1793 and now in the British
Museum.

assigned to the young Sidonius, and on January 1st he
delivered his address.[1] It contains 603 lines of hexameter verse and is preceded by a short elegiac introduction,[2] in which Sidonius modestly professes his
inferiority to his subject. There are many vices in this
work, and they are the vices that reappear in everything that Sidonius wrote—the paucity of ideas, the
obscuration of the main point, the laboured verbal
antitheses. The last, however, Sidonius evidently
considered a beauty; and it is only fair to admit that
in his hands they are sometimes effective. 'Ignotum
plus notus, Nile, per ortum'[3] is above his usual level,
and the poem also contains the fine phrase: 'has nobis
inter clades ac funera mundi / mors uixisse fuit',[4]
perhaps the only good lines that Sidonius ever wrote.

The whole poem owes much to Claudian: not only
are there many quotations[5] and reminiscences of him
throughout, but the whole scheme of the panegyric is
Claudianic: the divine machinery is brought out as
before, and all the hierarchy of heaven struts through
the verses of Sidonius. We have seen that part of his
education consisted in a mythological training, and in
this, as in many of his works, he was eager to display
his knowledge. The mythological disquisitions are the
worst part of the panegyric: at the beginning of it there
is a catalogue of the gods who attend the divine council
at the invitation of Jupiter:[6] twenty-one are mentioned,
each with an epithet or attribute; such lines as 'Iuno
grauis, prudens Pallas, turrita Cybebe' succeed each
other. This is very uninspired writing. It is worse than
mere padding, it is very bad padding. The epithets add

---

[1] *C.* vii. 7–10.   [2] *C.* vi.   [3] *C.* vii. 44.   [4] Ib. 537–8.
[5] See Geisler, *ap.* Lütjohann, pp. 395–9.   [6] *C.* vii. 17 *et seq.*

nothing to the picture and the gods' names hardly much more. Even the account of Rome's past[1], which is not pointless, is so over-elaborated that, though it is introduced to make a contrast, it fails in its effect and the point of contrast is lost. Though, as we should expect, the poet's emotions were never stirred by these pagan disquisitions, we can at least say that in later life he learnt how to handle such material less incongruously.

Yet, with all its faults, this poem, the first of his that is preserved, is in some ways the best that Sidonius ever wrote. The central figure was a man that he knew, who came from a country of which he was proud, and as a result the poem, when once the gods are out of the way, moves with real vigour. The career of Avitus is sketched in strong lines, and the poet really does bring it to a climax. Unfortunately in true Sidonian manner he does his best to spoil all by a ridiculous observation that the old emperor will rejuvenate the old Rome, which has been aged by his young predecessors.[2]

It is a notable fact about Sidonius' poetry, and indeed a commentary on his mind, that he could not create heroic figures. He could only write up to a noble subject, but he could not himself ennoble it. Thus Avitus, the essentially second-rate unromantic man, does not lose stature in the panegyric, while Aëtius and Attila, whom even the illiterate Jordanes could cast in heroic mould, pass like cold shadows across Sidonius' page. But when the grandeur of the subject was, as it were, a commonplace, Sidonius could write up to it. Such a subject he had to hand in the city of Rome. It is most remarkable how the name of Rome grows in stature as

[1] Ib. 50–122. [2] C. vii. 596–8.

her power declines. The authors of the later Empire,
provincials for the most part, seem to make her appear
greater as they stand farther away.[1] To them, the *Urbs-
orbis* assonance seemed almost designed to express a
mystic union between the city and the civilized world,
and to identify their fates.

Sidonius himself felt to the full this almost superstitious
reverence for the Roman name. It is Rome that is the
real hero of the panegyric: Rome, of whom ever since
her foundation it was law that she grew greater from
her disasters.[2] Her name was venerable:[3] since the
beginning of time there was nothing in the world better
than Rome, and nothing in Rome better than the
senate[4] of Rome. Alaric's one crime was her capture.[5]

It is not a little curious that, mingled with this glori-
fication of Rome, there is to be heard a very real note of
triumph, a proud announcement of the duty that Gaul
had taken upon herself. The saviour of the world was
coming from Gaul;[6] Gaul would do what Italy could
not, it would save Rome from the Vandals.[7] This pride
of a Gaul in his own land has led to something like
inconsistency in the poem: after the praise of Rome, on
which we have commented, it comes as rather a surprise to find a casual mention of the 'trepida urbs;[8] and,
if the senate was really the noblest part of Rome, it was
perhaps hardly tactful in a panegyric delivered to the
senate itself to point out that it had run away a short
time before.[9] It is hard indeed to capture the mind of
Sidonius or of the fifth-century Roman, but it is fair to

---

[1] Cf. Rutilius, 63 *et seq.*; Paulin. Pell., *Euch.*, 37.

[2] 'Aduersis sic Roma micat cui fixus ab ortu / ordo fuit creuisse malis', *C.*
vii. 6–7.   [3] *C.* vii. 501.   [4] Ib. 502–3.   [5] Ib. 505–6.

[6] Ib. 539, 'Orbis, Auite, salus'; ib. 153–4, 'sed cedere Auitum / dum tibi,
Roma, paro'.   [7] Ib. 588.   [8] Ib. 545.   [9] Ib. 450, 'exilium patrum'.

recognize in this contrast between particularist patriotism and loyalty to the empire, between reverence for
Rome and contempt for her weakness, an attitude of
mind which is typical, not only of the fifth century, but
of many centuries after. And of equal importance is
the fact that, though the contrast is so clearly expressed,
the poet is apparently unconscious of it. Only a young
man with a young man's tactlessness could have
pointed it so strongly, and never again is it presented
so clearly to us. But this contrast and its reconciliation
is a key to the political history of the West. We have
seen Sidonius as the last representative of the Roman
culture, and now he reveals himself as the forerunner
of the Middle Ages.

Yet one wonders what the senators of Rome really
thought of it all, and whether they winced at some of
the passages in it. We are told that the senate applauded,[1] and under the eye of the emperor and his
body-guard, we may believe that they did; and the
poet was certainly rewarded by the erection of a statue
in his honour in the Ulpian library.[2] His statue beside
that of Claudian, it was a notable distinction for a man
of twenty-three. This encouraged him to publish the
panegyric and preface: he accompanied it with a
modest dedication to Priscus Valerianus, who seems to
have been at that time praetorian prefect of Gaul.[3]

---

[1] C. viii. 9–10.

[2] C. viii. 8; E. ix. 16. 3, vv. 25–8; Marou has shown that the Scholae of the
Later Empire were situated in the forum of Trajan (*Mélanges de l'École française
de Rome*, 1932).

[3] C. viii., *inscr.*; cf. E. v. 10. 2; Borghesi, *Œuvres*, x, p. 740.

III
CONIURATIO MARCELLIANA

OF all the years of Sidonius' life there are none which are more obscure to us than 456–9, the end of the reign of Avitus and the beginning of that of Majorian. No letter is preserved to us which can be dated to that time; for our knowledge we are confined to the panegyric of Majorian,[1] the group of poems connected with it,[2] and one short allusion in a letter of more recent date.[3] It is idle to speculate why Sidonius has left us no letters from this period of his life, but we must regret the omission. That important political developments were occurring in Gaul we can have no doubt, but of the motives and even the precise events we must be prepared to rest for the most part in ignorance.

Avitus had come to Italy with a mandate to crush the Vandals, and he prepared to carry it out. An attempt was made to obtain the co-operation of the Eastern Empire, but it was not successful. Marcian did, it is true, receive an embassy from the West with kindness, and a contemporary Spanish chronicler could speak of the concord of the emperors and the 'unanimitas imperii',[4] but it seems that Avitus was not actually recognized as emperor by the Eastern Court.[5] As a result, he could count on no assistance from his colleague, and in fact received none.

His next move was to send an embassy to Geiseric threatening him with invasion if he did not keep the *foedus*.[6] This was probably bluff: Avitus was hardly in

[1] *C.* v. [2] *C.* iv, vi, and xiii. [3] *E.* i. 11. 6.
[4] Hydatius, 166, 169 (ii, p. 28).
[5] See Baynes in *J.R.S.*, xii (1922), p. 223.
[6] Priscus, fr. 24 (*F.H.G.*, iv, p. 102).

a position to invade Africa, and Geiseric knew it. He replied to the demand by sending a fleet of sixty ships to ravage Sicily. The relatively small number shows the opinion that Geiseric had of Avitus' strength. It was, in fact, too small. Avitus dispatched Ricimer with a fleet to Sicily, and he defeated the Vandals at Agrigentum. They retired to Corsica, and were there completely destroyed in a second battle.[1]

But Avitus' success only contributed to his ruin. A severe famine, due doubtless to the loss of the accustomed African corn-supply, broke out in Rome; and the populace forced him to dismiss his Gothic troops, that there might be less mouths to feed. He was compelled to yield; but the troops now demanded their pay, and to meet their demands he melted down such of the bronze statues in Rome as had escaped the Vandal search.[2] This anti-Roman action united senate and people against him, compelling him to leave the city[3] and retire to Gaul.[4] Meanwhile the victory over the Vandals had given renewed confidence to Ricimer and the Italian *foederati*: with Avitus discredited and Theodoric occupied in Spain, he and Majorian could now carry out the plan that they had already formed, it may be, after the death of Valentinian. They revolted against Avitus and defeated his Patrician Remistus, probably a Goth, at Ravenna (Sept. 17th).[5] Avitus at

---

[1] This is the inference drawn by Bury (i, p. 327) from *C.* ii. 367; Priscus, fr. 24, and Hydatius, 176, 177 (ii, p. 29).   [2] John Ant., fr. 202 (*F.H.G.*, iv, p. 616).

[3] I agree with Cantarelli (p. 61) that the expression used by Gregory of Tours, 'luxuriose agere uolens' (*Hist. Franc.*, ii. 10 (11)), probably refers to this. The aspersions cast on the character of Avitus by Gibbon (ed. Bury, iv, p. 14, n. 32) on the strength of this passage and ps.-Fredegar, iii. 7 have been refuted by Hodgkin (ii, pp. 393–5).   [4] Hydatius, 177 (ii, p. 29), John Ant., l. c.

[5] *Fast. Vind. Priores* and *Cont. Prosper.* (date), 579 (i, p. 304), Theophanes, *ad a.m. 5948.* The name is Gothic according to Schönfeld (*Wörterbuch der altgermanischen Personen- und Völkernamen,* p. 187), cf. Stein, p. 549.

Arles was endeavouring to enlist Gothic support for the reconquest of Italy, but as the Goths were occupied elsewhere, he met with no success.[1] As a last resort, he collected the remains of the Gallic army around Arles under his old general, Messianus,[2] and pushed forward across the Alps. At Placentia he encountered Ricimer and Majorian, and was completely defeated. Messianus was killed in the battle and Avitus was forced to retire into the town, where he soon surrendered (October 17th or 18th, 456). At the request of Eusebius, bishop of Milan, his life was spared and he was consecrated bishop of Placentia.[3] He seems to have been put to death soon afterwards; his remains were transported to the church of Brioude near his Arvernian home.[4]

Thus perished Avitus, and with him the attempt of Gaul to rule the empire. Avitus has his place in history, and his reign offers lessons of some interest to the student of the fifth-century empire. It has nothing to do with

---

[1] Hydatius, 183 (ii, p. 30), 'promisso Gothorum destitutus auxilio'.

[2] Cont. Prosper., 580 (i, p. 304), cf. C. vii. 426–7.

[3] Fast. Vind. Priores (Oct. 17th); Cont. Prosper. (Oct. 18th), 580 (i, p. 304); Victor Tonn. (ii, p. 186); Marius Avent. (ii, p. 232); Jord., Get., xiv. 240, John Ant., fr. 202 (F.H.G., iv, p. 616). Theophanes, ad a.m. 5948.

[4] The fate of Avitus is uncertain. John of Antioch (l.c.) says that he was defeated and compelled to flee to a church (τέμενος), but Majorian besieged him until he died either of hunger or strangulation. This account seems to contain a distorted version of his being appointed bishop. Chron. Gall., a. dxi, 628 (i, p. 664) says outright that he was put to death by Majorian. The Italian chronicles are silent, and Hydatius, 183 (ii, p. 30), has the mysterious words 'caret imperio, caret et uita'. Evagrius makes him die of plague (ii. 77). This might be the official version of his death put about by Ricimer and Majorian. That he endeavoured to return to Brioude and died on the way is the version of Greg. Tur. alone (Hist. Franc., ii. 11 (12)). But our best authorities state that he died at Placentia, and they should be followed. Gregory's account may be an attempt to reconcile the defeat of Avitus at Placentia with the existence of his tomb at Brioude. (On the tomb of Avitus see Chaix, i, p. 101, n. 4.) It is rather strange that the account of Gregory is so readily accepted by modern writers (see Cantarelli, p. 62, n. 1).

separatism;[1] any such idea is inconsistent not only with what Avitus did, but with the actual documents, if carefully studied. To a Gaul Avitus was 'orbis salus',[2] and that alone refutes any view that he represented an idea of Gallic separatism. If he had, he might not have failed. It is interesting to speculate what might have happened if Avitus had confined his aim to making himself an emperor of Gaul, supported by the Gothic federates and with the Council of Arles elevated into a Senate equal with that of Rome. In that event, the empire might have split into three, the Gallic prefecture, Italy, and the East, each supported by federate barbarians. Avitus represented the homogeneous spirit of the Roman population of the West, and he failed primarily because he came into opposition with the naturally particularist feelings of the barbarian federates.[3] It is not Avitus but Ricimer who represents the disruptive tendencies of the period. Avitus, the man of simple mind, was a conservative trying to build up a united West upon the basis of Visigothic support. He failed because his attempt was out of touch with the conditions of his time. The opposition of senate and people in 456 seems not to have been opposition to the man because he was a Gaul so much as opposition to his measures because they were unpopular. It is true that in the next year Avitus' consulship seems not to have been recognized[4] and that Majorian could

---

[1] Avitus is considered as a portent of Gallic separatism by Camille Jullian (*De la Gaule à la France*, p. 216), and G. Tammasia (*Egidio e Siagrio*, pp. 10–11); see on the other side Duval-Arnould, pp. 15–18.

[2] *C.* vii. 339.

[3] See Seeck, *Untergang*, vi, pp. 335–7.

[4] De Rossi, i. 798 and 799 are dated 'post consulatum Iohannis et Varanae', and these were the two consuls recognized in the East for 456 (see Seeck, *Regesten*, p. 402).

congratulate the senate on being liberated 'a domestica clade'.[1] Still it is to be noted that an inscription from Rome dating from November 1st, 456, cites the consulship of 'Dominus noster Auitus'.[2] The statue of Sidonius remained upright in Trajan's forum after his master had been dethroned. That statue is significant. Its survival shows how little national ill-feeling there can have been between Italy and Gaul. The key to the history of this time is the opposition between the centripetal feelings of the imperial subjects expressed for example in the triumphant supremacy of the papacy, and the individualist aspirations of the military commanders, for the most part non-Catholics. The persistence of the universal rule of the Roman Church along with the rise and fall of separate kingdoms carved out of the Roman provinces is a result of that opposition which is represented on a smaller scale in the attempt and fall of the emperor Avitus.

We saw Sidonius last in Rome on January 1st, 456; he appears again at Lyons towards the end of 458, but the intervening events of his career are all darkness to us. At what date he left Italy, whether he took part (as has been supposed) in the final battle of Placentia, we cannot say.[3]

According to strict constitutional law there was no such thing as an interregnum after the death of Avitus.[4] If the emperor of one half of the empire died, the survivor automatically became emperor of the whole, but in practice Sidonius would not be wrong in referring to

---

[1] *Novell. Maioriani*, i. 3.

[2] De Rossi, i. 795. Yet it would seem that Avitus had been already deposed a fortnight before. Either the dating of the chronicles is wrong or news travelled very slowly in Italy at this period.

[3] Coville's hypothesis (pp. 56–7) for these years will be examined later. (Appendix A.)

[4] See Bury, i, pp. 16–18.

the situation as 'hiantis interregni rima'.[1] The *de facto* ruler in the West was Ricimer, the *magister militum praesentalis*. Such a situation was not at all to the liking of Theodoric. He had been campaigning with success against the Suevi in Spain on behalf of Avitus and the empire.[2] But on the news of Avitus' defeat, he returned with the greater part of his army to Gaul, doubtless to secure the territory assigned to the Goths in Aquitania, and to make a plan for the future. At the same time the Burgundian troops who had joined him in the Spanish campaign, perhaps as *foederati* of the empire, returned to occupy their old settlement in Savoy.[3] No less decisive was the effect of Avitus' overthrow upon the Gallo-Roman nobility. His defeat and death signified not only that their attempt to give a ruler to the Empire had broken down, but that, for all practical purposes, there was no emperor of the West at all. It seemed that the Gauls must submit to the rule of a barbarian master of soldiers who commanded the armies of Italy. Such a situation had never occurred before, nor is it surprising that many of the Gallo-Romans refused to accept it, and resolved to create upon their own initiative a successor to Avitus. Unfortunately the details of their project are matters of the greatest obscurity. Their action was called by Sidonius at a later date the 'coniuratio Marcelliana', or so it appears in the best of his manuscripts.[4] Its inception seems to

---

[1] *E.* i. 11. 6.

[2] 'Fidus imperio Romano' is the phrase used by Hydatius, 170 (ii, p. 28).

[3] Burgundians under Gundioc and Hilperic in Spain, see Jord., *Get.*, xliv. 231. Cf. *Cont. Prosper.* (i, p. 305) and Hydatius, 186 (ii, p. 30) (Gothic troops still remaining in Spain after the return of Theodoric).

[4] According to Mohr's *apparatus criticus* the reading 'Marcelliana' is given by the *Matritensis* (Mohr's C.), the others have the non-existent form 'Marcellana' (pp. xiv, 24; cf. p. 376).

date from late in 456.[1] It has usually been assumed that the object of the conspiracy was to raise to the empire Marcellinus, who was at this time ruling independently of both empires in the province of Dalmatia. This may be the correct interpretation, and it is not improbable in itself. Still it is only fair to say that no manuscript gives 'Marcilliniana', the correct adjectival form from Marcellinus, and in view of the lack of any contemporary letters of Sidonius, we cannot say that, if there had been at this time a Marcellus around whom a conspiracy was woven, we must have heard of him elsewhere. After all, he may well have met his death before the full stream of Sidonius' correspondence begins; that is often the fate of conspirators.[2] It is certain that the Gauls counted on a measure of barbarian support for what was in fact an attack upon the pre-eminent position of Ricimer and the Italian army, and it is highly probable that a mysterious entry in Marius of Aventicum, who states that the Burgundians occupied a part of Gaul and divided the lands with the Gallic senators, refers to some sort of pact between the conspirators and them.[3] Moreover, it was not long

---

[1] The praefecture of Paeonius, which lasted a full year, commenced at a time when there was an interregnum ('uacante aula'), i.e. Paeonius became praefect at the beginning of 457. It appears from the narrative of Sidonius that his tenure of the prefecture was subsequent to the outbreak of the conspiracy (E. i. 11. 6).

[2] There was a Marcellus who was praetorian prefect of the Gauls in 444-5 (Borghesi, Œuvres, x, p. 736; C.I.L., xii. 5336) and a Marcellinus of Narbo, a contemporary of Sidonius (E. ii. 13. 1; C. xxiii. 463-5). The view that the 'coniuratio Marcelliana' concerned Marcellinus of Dalmatia has been almost universally adopted, Dalton alone expressing a doubt (i, p. xx, n. 3, and ii, p. 221). The support for it so often cited from Procopius, Bell. Vand., i. 6. (cf. Mommsen, ap. Lütjohann, p. 430) is seen on close examination to be no support at all. Allard (p. 48, n. 2), however, knows better. 'Procope dit qu'il fut "roi de Lyon" (Λέων βασιλεύς, i.e. the emperor Leo!!). 'Il est probable', he adds, 'que Marcellin n'était pas alors à Lyon.' That at least is true.

[3] Marius Avent., ii, p. 232. Cf. Cont. Prosper., i, p. 305. The wildly con-

before a barbarian garrison appeared in Lyons.[1] As Theodoric is said to have acquiesced in the action of the Burgundians, it may be inferred that he also gave his support to the 'coniuratio Marcelliana'.[2] Among its leaders was a certain Paeonius, who, though a *nouus homo*, and only of the rank of *spectabilis*, took advantage of the interregnum to seize upon the office of praetorian prefect of Gaul.[3] Whether Sidonius was a party to this movement has been much debated, and certainty is impossible to reach. We can at least say that as a land-owner in *Lugdunensis* he could be considered responsible in strict justice for the Burgundian settlement, whether he had approved of it or not; and from the whole tenor of his life and activities it is hard to believe that he could have acquiesced in a state of affairs which left the West without an emperor to the caprices of barbarian commanders.[4]

As has been suggested, the conspiracy was directed chiefly against the autocracy of Ricimer, and when Majorian was declared emperor (April ?, 457)[5], the

fused narrative of ps.-Fredegar, ii. 46, is out of its chronological order and its text is not certain. Schmidt (i, p. 255) indeed gives it up in despair, but it seems to contain a grain of truth in the statement that the Gauls assigned lands to the Burgundians in *Lugdunensis*.

[1] On the barbarian garrison see *C.* v. 571–3; ps.-Eusebius, *Homilia de Litaniis* (de la Bigne, *Bibliotheca Maxima Patrum*, vi, p. 652); Schmidt, i, p. 373. The language of *C.* v. 572, 'uisceribus miseris insertum *depulit* ensem' [not 'hostem', as in Sirmond], is against Coville's theory of a Roman garrison placed there in 458 and subsequently withdrawn (p. 130).

[2] *Cont. Prosper.*, i, p. 305.

[3] Allard (p. 49, n. 6) believes that Paeonius was *vicarius*, on the ground that he was only 'uir spectabilis'. But he is expressly called by Sidonius 'uir praefectus' and 'uir praefectorius'. That a *praefectus praetorio* was *uir illustris* as a rule is true, but the whole point here is that Paeonius became *praefectus praetorio* under exceptional circumstances, and did not until later obtain the *codicilli* from the emperor (cf. *E.* i. 11. 6 and 7).

[4] See Appendix A.

[5] The difficulties connected with the chronology of Majorian's accession are

greater part of its *raison d'être*, as far as the Gallo-Romans were concerned, had disappeared. Paeonius, who was confirmed in his office towards the end of 457, received at last the proper title of *illustris* and with it the *codicilli*, that marked his official recognition;[1] from this we may infer that he had already abandoned his support of the conspiracy. Magnus, one of the foremost nobles of south Gaul, about this time declared his acceptance of Majorian.[2]

But there were others, who, whether they liked it or no, were too deeply compromised to escape from the consequences of their actions. There were the inhabitants of *Lugdunensis prima*, Sidonius among them, who had admitted the barbarians into their midst, and, as always, found barbarians much easier to invite than to dislodge. Many of them, very possibly, were merely victims of circumstances and had no ill-will against Majorian; though in Lyons, where Avitus' family had owned lands,[3] there may well have been an opposition on personal grounds to the new emperor which does not, at least as far as our meagre evidence goes, appear to have existed with such intensity in south Gaul. It is significant that in an inscription of June 458, Majorian seems not to have been recognized as consul.[4]

discussed by Stein (pp. 553–4) and Baynes in *J.R.S.*, xii (1922), pp. 233–4, and xviii (1928), pp. 224–5. The questions cannot be regarded as settled yet, but it looks as if Baynes has the better of the argument.   [1] *E.* i. 11. 6.

[2] The Gallic attitude towards Majorian and the rationale of the 'coniuratio Marcelliana' are illustrated by the certainly disingenuous, but not, as I see it, entirely false picture given by Sidonius in *C.* v. 9–10: 'fateor, trepidauerat orbis, / dum non uis uicisse tibi', &c. If the 'coniuratio Marcelliana' is really connected with Marcellinus of Dalmatia, it is significant that about this time he promised support to Majorian (Priscus, fr. 29 (*F.H.G.*, iv, p. 103)). I accept from Sundwall (p. 98) the view that *C.* v. 558 and *C.* xv. 154–7 refer to Magnus, but I should assume that his appointment as *magister officiorum* in Spain dates rather from Avitus' reign.   [3] *E.* ii. 12.

[4] Allmer et Dissard, iv, p. 28; *C.I.L.*, xiii. 2363. Mommsen (in *C.I.L.*, xiii,

Yet, however strong may have been the feeling against Majorian, he seems to have taken no action against the insurgents for a considerable time: possibly he did not regard them as of great moment, for, if we may trust the order of events in Sidonius' panegyric, a contribution for the proposed African expedition was imposed on the Gauls, before Majorian had crossed the Alps.[1] In any case, his attention was sufficiently occupied by Vandal ravages on the Italian coast in 457, and an invasion of the Huns in 458.[2]

Some time towards the end of 458, it seems that Petrus, the *magister epistularum*,[3] was sent over the Alps with an army against Lyons.[4] After a stubborn siege, in which much damage was done to the town, the Burgundians completely submitted: they were forced to abandon Lyons and become *foederati* of the empire.[5] The inhabitants were compelled to give hostages and the usual capitation tax was increased threefold.[6]

p. 365) denies that the inscription can date to 458, because, if so, Majorian's name could not be omitted. But that is to forget the history of the period.

[1] *C.* v. 446–8.

[2] The series of Majorian's laws shows that he did not leave the neighbourhood of Ravenna until Nov. 458.

[3] 'Petri . . . magistri epistularum', *E.* ix. 13. 4; 'qui scrinia sacra gubernat', *C.* v. 564. On Petrus' campaign against Lyons see *C.* v. 571–84. The date is roughly fixed from 'nuper' in ib. 571.

[4] That Petrus was sent in advance by Majorian and captured Lyons before his arrival is a certain inference from *C.* v. 584–5, 'ueniens [addressed to Majorian] tamen omnia tecum / restituis'. This was first pointed out by Kaufmann in *Neues schweizerisches Museum*, 1865, p. 7, but his date (end of 457 or beginning of 458) is too early. Stein's theory that the capture of Lyons was due to Aegidius operating from the north is ingenious (i, p. 559), but seems to deny the role clearly assigned by Sidonius to Petrus. Moreover, the evidence connecting Aegidius with north Gaul at this time is very weak and in *Lib. Hist. Franc.*, 8 a confusion between Aegidius and Aëtius may be suspected: there is a rather similar confusion in Gildas, *De Excidio*, 20.   [5] See Schmidt, i, p. 257.

[6] The interpretation adopted is that of Schmidt (i, p. 256); cf. F. Lot in *Revue historique du Droit*, 1925, p. 32, where the passage is somewhat differently explained.

It was fortunate for Sidonius that he found in Petrus a kindred spirit. He was a poet of considerable repute[1] and it was probably at his instigation that Sidonius determined to make his peace with Majorian. The emperor had crossed the Alps at the end of 458 in the depths of winter,[2] resolved to settle the question of the Burgundians and to force upon the Visigoths a renewal of the *foedus*. Sidonius at once went to meet him, and delivered to him a short poem, in which he prayed for alleviation of the fine imposed upon him. The Hercules of his age, declared Sidonius, will be able to cut off the three heads from the hydra of my capitation tax.[3] It seems that his prayer was successful; and when later in the year Majorian arrived at Lyons, Sidonius was induced by Petrus to deliver before him a public address, the object of which was to obtain for the exhausted population of Lyons the favour which had been granted to himself (end of 458)[4].

This address, the *Panegyricus Maioriani*, is a poem of 603 hexameter verses, one verse longer than the *Panegyricus Aviti*, and, like it, preceded by a short introduction in elegiacs. From the circumstances it is obvious that it must have been composed in considerable haste, and it says something for Sidonius' technical skill that it does not show many signs of it.[5] Taken as a whole

[1] See *E.* ix. 13. 5, vv. 6–8, 76–87; *C.* ix. 308.

[2] After Nov. 6th (*Novell. Maioriani,* vii). For the crossing of the Alps, *C.* v. 510–52.

[3] *C.* xiii. For the chronology of *C.* iii. iv. v. and xiii, and the circumstances of their delivery see Appendix A.

[4] As Cantarelli (p. 71, n. 4) says, *C.* v. 2, compared with ib. 378 look as though they ought to suggest a date early in 458, but the series of Majorian's Novels from Ravenna show that it is inadmissible. See also Mommsen (*ap.* Lütjohann, p. li).

[5] The roughness of the transition at verse 327 is particularly noticeable, and indeed is so marked that possibly there is corruption in the text. In v. 352

the poem is inferior to the panegyric of Avitus. Sidonius was not so well acquainted with the life-history of
his second hero, and consequently, as the poems are of
practically the same length, there is more mythological
padding. It is certainly introduced with greater relevance, and there is an increased command of the
material which seems to indicate the more mature
writer: there is nothing like the catalogue of gods,
which is a feature of the earlier poem. On the other
hand, the necessity of writing a long poem without
having much to say has tempted the poet to display
at times his historical and mythological knowledge to
most incongruous effect. Thus nearly a quarter of the
poem is taken up with a speech in which the wife of
Aëtius inveighs with jealous envy against the rising
Majorian.[1] That Aëtius' wife should be represented as
jealous has value in a panegyric, but not to the extent
of a hundred and thirty lines, and the speech has an
intrinsic vice. Sidonius describes the woman as 'bar-
bara',[2] and any competent verse-maker should have
been able to write a hundred or so of forcible lines
appropriate to a jealous barbarian woman: Sidonius
was perfectly able to do this, but he was too lazy or too
hurried to try. Instead, the barbarian woman exposes
her knowledge of Sicilian Eryx, Amycus, Hippome-
nes, Hector, Ulysses, Bacchus, Centaurs and Lapi-
thae, Theseus, and many other Grecian heroes, of
whose names she was in reality completely ignorant, as
Sidonius well knew. That artistic economy of force
should be sacrificed to a desire to exploit the knowledge
of facts was a convention of the times, and Sidonius

Rome tells Africa who her saviour is to be, but in v. 104 Africa is represented as
already knowing his name.　　　[1] *C.* v. 143–274.　　　[2] Ib. 128.

cannot be blamed for conforming to it, but writing such as this is nothing short of artistic dishonesty.

Though the poem as a cohesive whole falls below the standard of its predecessor, in many of its details it surpasses it. The personification of Rome, in spite of its somewhat over-elaborate ornament, is better worked out, and the confrontation of Africa and Rome has real dramatic value, of which the poet has made good use. He should not be judged as a master of feeling and emotion, but as a rhetorician who happened to possess the technical knowledge of versifying. As a rhetorician, we may think him generally unsuccessful, but even to our ears he is not entirely despicable. For an indictment on the fifth-century bureaucracy—'pretium res publica forti / rettulit inuidiam'[1]—has a terse grandeur, and the noble phrase—'tua nempe putantur / surgere fata malis et celsior esse ruina'[2]—is not unworthy of the Roman tradition. The lines describing the miserable condition of Lyons prove that Sidonius could communicate in some degree the depths of his own emotion, but they also show how the rhetorical tradition could bring out not only his best but his worst. The fifteen vigorous lines[3] are finished with a climax the falsity of which is nothing but ludicrous;[4] no malevolent critic could hurt Sidonius so much as he harms his own sincerity and sense by allowing his rhetoric to get out of hand.

The best passages in the panegyric are those which describe the barbarians. It is remarkable that the Franks are portrayed with much more vigour in this poem than the Visigoths had been in the earlier pane-

---

[1] 362–3.   [2] 63–4.   [3] 571–86.

[4] 'fuimus uestri quia causa triumphi, / ipsa ruina placet' (We are happy in our ruin, for it is the cause of your triumph).

gyric. One cannot help feeling that in the interval
Sidonius had mixed much with barbarians, and possibly
as a Lyons landowner he had for the first time talked
with a barbarian warrior. He did not like barbarians,
and in his letters he expressed that dislike in uncompromising tones. He disliked them all, he tells us, even the
good ones.[1] Yet in this poem he seems to find in them
a *divinum aliquid*, and accords to them an admiration the
more remarkable because it is reluctant. The courage,
the pertinacity, and the individualism of barbarians
are described with an almost sympathetic understanding. Even the typically Sidonian pun for once
has dignity when he tells us of Frankish courage:
'si forte premantur, / seu numero, seu sorte loci, mors
obruit illos, / non timor; inuicti perstant, animoque
supersunt / iam prope post animam'.[2] True and faithful supporter of the Roman tradition as he ever was,
there seem moments when, in spite of himself, Sidonius
stretches out his hands towards a farther shore.

Not only is the panegyric of value as an historical
document and as a milestone in the development of
Sidonius' technique, it has also an ethical importance
for an examination of the author's own character.
That Sidonius should deliver a laudation of the very
man who had taken part in the deposition, and quite
possibly the death, of his father-in-law has seemed in the
eyes of many to indicate a lack of earnest feeling and a
reprehensible compliance of spirit.[3] It is easy to con-
demn, and many who have admired Sidonius' bold stand
for Church and Empire in later years may regret that
the panegyric was ever delivered. Yet he might have

---

[1] *E.* vii. 14. 10; cf. *E.* iv. 1. 4; iv. 8. 5; *C.* xii. [2] *C.* v. 250–3.
[3] As, for instance, Dalton, i, p. xxiii.

pleaded that in the events which led up to the death of
Avitus, Majorian took only a secondary part. Majorian
was after all a true Roman, and he was no stranger to
Gaul[1] where his victories in the stormy years 430–40
had laid the foundation for that quiet social life which
is a feature of Sidonius' correspondence. If he had
been carried away by the genuine greatness of his subject, it would be easy to pardon him. But that was not
all: Lyons, his own birthplace, was in real danger. It had
opposed the reigning emperor, and had been punished
with a heavy fine; hostages had been taken from it.
Would not a man feel justified in doing almost anything
to save the town in which he had been born? It is the
verses lamenting the pitiful condition of Lyons which are
the excuse for the panegyric, and the author had a right
to plead that the excuse was valid. Whether Sidonius
was successful in gaining his point we cannot say with
certainty. At least he felt himself justified shortly after
in publishing the Panegyric and Preface with a short
dedication to Petrus, the Maecenas of this time.[2]

The remainder of Majorian's Gallic campaign is soon
told. He advanced from Lyons against Theodoric, who
was besieging his general, Aegidius, in Arles,[3] as it
appears, and the Gothic king, with a part of his army

---

[1] *C.* v. 207 et seq. Allard (p. 54) has no authority for saying that the father
of Majorian 'avait rempli en Gaule un haut emploi de trésorerie', nor that the
emperor himself was 'né ou au moins élevé dans les Gaules' (p. 60).

[2] *C.* iii. 5.

[3] Paulinus Petrocordiensis, vii, 111–51; Greg. Tur., *de Mirac. S. Martini*, i. 2.
The incident which they record was first dated by Tammasia (p. 12). The river
Rhone, a bridge of boats, and an 'urbs' are the only topographical details given,
but we cannot fix Aegidius on the Rhone at any other time. The incident has
been referred since Dubos (*Sur l'Établissement de la monarchie françoise*, ii, p. 56)
to Arles, where there was certainly a bridge of boats in the fourth century
(Ausonius, *Urbes*, 77, 'Pons naualis'; Cass., *Var.*, viii. 10. 6). Vienne is also a
possibility (cf. Jullian, *Hist. de la Gaule*, iv, p. 119).

employed in Spain, was unable to make much resistance:[1] after suffering several defeats, he was compelled
to renew the *foedus*.

Majorian seems to have remained for a year at Arles,
completing, as we may suppose, the preparations for his
expedition against Geiseric.[2] During some of this time
Sidonius too, at the invitation of the emperor, was staying in the town, and to some date in these months is
probably to be referred the supper party which many
years after he recalled as having taken place 'in the
times of Majorian the Emperor'.[3] Petrus had just
published a book in prose and verse, and Sidonius, with
Domnulus, Severianus, and Lampridius, three literary
men of the time, had been invited to a banquet in his
honour. Each was to produce a poem celebrating the
new book. Sidonius' own composition, an Anacreontic
in seventy-six verses, was not included by him in his
collected poems, and in sending a copy of it to a friend
some years later he pointed out that it was not a serious
work and should therefore be exempted from serious
criticism.[4] It is a graceful and pleasantly flowing piece
which proves at least the author's facility. Many of
Sidonius' more pretentious pieces have no more depth
of feeling than this slight effusion, and the attention
that he gave them did little more than rob them of the

---

[1] Hydatius, 197 (ii, p. 31); Priscus, fr. 26, (*F.H.G.*, iv, p. 103). *C.* v. 562-3
refers, as Schmidt has shown (i, p. 257), to an unsuccessful attempt to treat with
Theodoric at the end of 468.

[2] See *Novell. Maioriani*, x and xi; Cantarelli, pp. 73-4.

[3] *E.* ix. 13. 4. The incident is dated by Baret (p. 145) and Dalton (ii, p. 251)
to 461, wrongly, as I believe. The work of Petrus has already been acclaimed
in Italy and Gaul (*E.* ix. 13. 5, vv. 104-14, 'dat . . . resultat'), but its reception
in Spain is placed in the future, 'imitabiturque Gallos / feritas Hibericorum',
ib., vv. 115-16). From this it appears probable that the banquet preceded the
Spanish expedition of Majorian.

[4] Ib.; cf. § 5.

merit of spontaneity. Kindly critics will not take Sidonius at his word and leave these table-verses out of account in making an estimate of his poetry.

At what precise date Sidonius left Arles we do not know: by the summer of next year he was in Auvergne, and it is probable that, during this period of his residence in his adopted country, he held by favour of the emperor the important title of *comes ciuitatis Arvernorum.*[1] For a man who still considered himself as young, he was making good progress along the path of public distinction that his ancestors had trod before him.

It is unfortunate that no details of Sidonius' administration in Auvergne are given to us: no letter exists which can be assigned to this period of his life. It was not long, however, before he was again at Arles, and the circumstances of his visit are so curious that they deserve to be related at some length. In the letter in which Sidonius describes it we are given, indeed, our last glimpse of the personality and private life of a Western emperor.

In 461 Majorian, disgusted and disillusioned, it may be, at the failure of his great expedition against the Vandals, was returning to Italy. On his way he halted at Arles,[2] and was entertained with games at the amphitheatre.[3] During his stay a satire was circulated

[1] In 461 we see (i) that Sidonius has just come to Arles from Auvergne (*E.* i. 11. 4, 7), (ii) that he is still 'militans' (ib., § 1), (iii) that he is addressed by Majorian as 'comes Sidoni' (ib., § 14). By far the simplest explanation of these three passages is to suppose as Savaron (*Origines de la Ville de Clairmont* (1662), p. 34) long ago suspected, that Sidonius was serving in Auvergne as *comes ciuitatis Arvernorum.* On this office see Seeck (in *P.–W.*, 'Comes') and Fustel de Coulanges, *L'Invasion germanique*, pp. 49–59. Sidonius mentions a *comes ciuitatis* of Marseilles, *E.* vii. 2. 5 (cf. *E.* v. 18). As the 'commilitium' of *E.* i. 11. 3 involved 'peregrinatio' and is described as 'recens' it cannot be identified with that of § 1. Its drift has been explained above (p. 49).

[2] *Chron. Gall.*, a. dxi, 633, 635 (i, p. 664).   [3] *E.* i. 11. 10.

in the town, holding up to ridicule many of the prominent citizens.[1] It was a violent piece of work, and
created a great sensation among the inhabitants.
Paeonius, whose doubtful ancestry and unexpected
rise to power we have already noted, was spared least
of all. The satire was anonymous and all were wonder-
ing who was the author. Sidonius had enemies in
Arles, Paeonius among them, who were jealous of the
friendship which he had formed with the emperor, and
angry that one who had taken part in the Marcellian
conspiracy should have escaped so lightly.[2] It was not
difficult to ascribe the authorship to him, especially as
he was now a poet of established reputation; and when
the Arvernian Catullinus, who was on a visit to Arles,
expressed a loud and scarcely discreet approval of it,
Paeonius was sure that he had found the author.
'Watch Catullinus there, bursting with laughter,' he
cried. 'He would not laugh at parts of the satire in this
way unless he had heard it all. Sidonius is in Auvergne;
it is obvious that he wrote the satire and that Catullinus
heard it from him.' As the work was libellous, the
author was liable to punishment,[3] and if it was known
that a *comes ciuitatis* had written it, matters would
obviously be worse. Suspecting nothing of the intrigues
around him, Sidonius appeared in Arles, and after
paying his duty to the emperor went down to the

---

[1] 'Satira ... uersuum plena satiricorum mordacium', ib. § 1, 2. The incidents
described in *E.* i. 11 are clearly dated to 461 from ib. 10, 'Magnus ... olim
ex praefecto, nuper ex consule' (Magnus had been consul in 460, see Liebenam,
Fasti, p. 47).

[2] 'quamquam putarer ab *inimicis* non affuturus', *E.* i. 11. 7. Cf. ib. § 2, 'quid
fuerit illud, quod me sinistrae rumor ac fumus opinionis afflauit ... exponam'.

[3] See *Cod. Theod.*, ix. 34. A *famosus libellus* was defined by Ulpian (*Dig.*, xlvii.
10. 6. 3) in language that certainly covers this case. The extreme punishment
was death, which might also be inflicted on any one who did not destroy the
work if it came into his hands.

Forum. He could at once see that something very strange was afoot. Men approached him with exaggerated reverence, bowed, and passed on; men walked beside him with sullen glances; others, to avoid paying their respects to him, were skulking behind columns and statues. One of the intriguers advanced towards him. 'Do you see these men?' he said. 'Yes,' replied Sidonius, 'I do, and their behaviour fills me more with astonishment than with admiration.'[1] 'Well,' said he, 'they take you for the author of the satire, and are accordingly either annoyed or alarmed at you.' 'Satire!' cried Sidonius. 'Where? Why? When? Who says so? Kindly ask your friends whether the informer who pretends that I have written satire alleges that he has seen my handwriting.' At this, according to Sidonius, his foes appeared to be satisfied.

On the next day Sidonius was invited to take dinner with the emperor.[2] The company included his kinsman Magnus, the ex-prefect, who had been consul in 460, Camillus, Magnus' nephew,[3] and Paeonius. When the meal was nearly finished Majorian opened the conversation: after a few remarks directed to Severinus, the consul of the year, he engaged Magnus in a short talk on literary subjects. Camillus' turn came next. 'You have an uncle, brother Camillus' [i.e. Magnus], said the emperor, 'who makes me proud that I have conferred a consulship on your family.' 'Not one, Augustus, but the first,' was the quick reply of Camillus, who hoped for a similar honour for himself. Not even the presence of the emperor could check the round of applause at the ready answer. Paeonius was next in

---

[1] 'Et ego "Video", inquam, "gestusque eorum miror equidem nec admiror",' E. i. 11. 8.   [2] Ib. § 10–17.   [3] Camillus, cf. C. ix. 8.

order at the table, but either from accident or design, Majorian passed him over and addressed a question to his neighbouring Athenius. Paeonius rudely answered the emperor before Athenius had time to open his mouth, but Majorian only laughed, 'for,' says Sidonius, 'while maintaining his dignity he liked a joke in company'.[1] The crafty Athenius, his annoyance calmed by the merriment of the emperor, observed: 'I am not surprised that Paeonius should try to push himself into my place, when he has pushed himself into your conversation.' The next guest, Gratianensis, remarked that the incident would provide good material for a satirist. The emperor now turned to Sidonius. 'I learn, Count Sidonius,' he said, 'that you are the author of a satire.' 'I have just learnt it too,' replied Sidonius. 'Well, spare us, at least.' 'In refraining from illegality I spare myself.' 'And what shall we do,' said Majorian, 'to your accusers?' 'Whoever they may be,' replied Sidonius, 'let them come out into the open. If they prove their case, I am willing to suffer the full penalty: if, as will not improbably be the case, I refute the charge, I beg of your clemency to give me leave to write what I wish consistent with the law against my accuser.' 'I agree,' said Majorian, 'if you do it here and now.' It was a hard request, but Sidonius did what he could and in an instant had composed a distich.

> ' Who says I write satires, great lord,
>   Must prove it, or pay for the word.'[2]

Great applause greeted the effort, though it was

---

[1] 'Subrisit Augustus, ut erat auctoritate seruata, cum se communioni dedisset, ioci plenus', *E.* i. 11. 12. Gibbon well says that these words of Sidonius 'outweigh the six hundred lines of his venal panegyric' (ed. Bury, iv, p. 26, n. 65).

[2] 'scribere me satiram qui culpat, maxime princeps, / hanc rogo decernas aut probet aut timeat'.

earned, as Sidonius suggests with a modesty in which
one will agree with him, rather by the speed than by
the merit of the production. The emperor in a speech
of perhaps deliberate solemnity declared him completely innocent.[1] Soon after this the company broke
up, but Sidonius could not resist making a last attack
on Paeonius. He would never satirize Paeonius' intrigue, he explained to the departing company, so long as
his own actions were not in future misrepresented. For
it was quite enough for him that the ascription of the
satire had brought credit to him and infamy to his
rival.[2]

Anecdotes from the life of the fifth century are not
very common, and for that reason this one has been
treated at some length. The whole incident is unim-
portant enough, and, when we are introduced to
Majorian at a banquet, we can only feel disappointed
that, while he might have said so much that would be
interesting, the conversation is really so trivial. As an
illustration, however, of the character of Sidonius, it has
a very real value. The scenes in the Forum and at the
banquet are quite typical of the man and his child-like
desire to score off his opponents, combined with his
determination to preserve his dignity and gravity in
doing so. The pompous speeches of § 13 and § 15 and
the naïveté with which the whole story is told are perfect
indications of the man Sidonius; and we may note also
the delightful way in which he tells his correspondent
how satisfied he felt to see all the officials and digni-
taries bowing before him.[3] Best of all, perhaps, is the

---

[1] It almost seems as if the emperor were deliberately parodying Sidonius'
own style of talking, and we need not be surprised if Sidonius failed to see this
joke against himself.   [2] E. i. 11. 16.

[3] Ib. 17, 'sed cum mihi sic satisfactum est, ut pectori meo pro reatu eius tot

ponderous speech of § 8 with its attempts at being forcible which only lead to its being incoherent, and its inevitable Sidonian word-play. In his whole works, there is nothing more delicious.[1] Sidonius had indeed the true aristocratic dignity, but, whether he was talking or writing, he never learnt that other aristocratic virtue of knowing when he had said enough.

potestatum dignitatumque culmina et iura summitterentur, fateor exordium contumeliae talis tanti fuisse, cui finis gloria fuit'.

[1] 'Perge, amice, nisi molestum est, et tumescentes nomine meo consulere dignare, utramnam ille delator aut index, qui satiram me scripsisse confinxit, et perscripsisse confinxerit: unde forte sit tutius, si retractabunt, ut superbire desistant.'

IV

THE PERIOD OF RETIREMENT (461–7)

I T is sad to reflect that as he sat at the banquet of
Arles with a joke for all, Majorian was a doomed
man. His creator, Ricimer, had found him too strong
a man, and too capable a military leader: he felt that
the emperor might at some time remove him from his
supreme position as commander of the Italian *foederati*.
Conscious of his danger, he resolved to take advantage
of the fact that Majorian, unlike Avitus, was not
accompanied on his journey to Italy by a Visigothic
body-guard.[1] Hardly had the emperor crossed the
Alps, when he was met at Dertona by Ricimer and
ordered to resign the purple (August 2nd, 461).[2] Five
days later he was dead. It was given out that he had
died of disease:[3] the same story, we may remember,
had been circulated about his predecessor, Avitus, but
contemporaries did not doubt that he had been murdered.[4] He was succeeded on the throne by Libius
Severus, a nominee of Ricimer.[5]

---

[1] John Ant., fr. 203 (*F.H.G.*, iv, p. 616). "Ὁ μὲν γὰρ [Μαιουρῖνος] τοὺς συμμάχους
μετὰ τὴν ἐπάνοδον ἀποπέμψας, σὺν τοῖς οἰκείοις ἐπὶ τὴν Ῥώμην ἐπανήρχετο."
We are ignorant of the motives of this apparently suicidal act. Perhaps Majorian
was really quite unsuspecting. Jordanes seems to refer to a campaign projected
in this year against the Alani (*Get.*, xlv. 236).

[2] *Fast. Vind. Priores*, 588 (i, p. 305) (date); John Ant., fr. 203 (*F.H.G.*, iv,
p. 616).

[3] Procopius, *Bell. Vand.*, i. 7. 14, and see below, next note.

[4] *Fast. Vind. Priores*, 588 (i, p. 305) (date); *Chron. Gall.*, a. dxi, 635 (i, p. 664);
Hydatius, 210 (ii, p. 32); Cassiodorus, *Chron.*, 1274 (ii, p. 157); John Ant., fr.
203 (*F.H.G.*, iv, p. 616); Cantarelli, p. 76. Theophanes (*ad a.m. 5955* and *a.m.
5964*) gives the two different versions of the story under two different dates—
both wrong.

[5] If July 7th, the date given for the accession of Libius Severus by Victor
Tonnennensis, *ad. ann. 463* (ii, p. 187), and Theophanes (*ad a.m. 5955*) is correct,
it would mean that he was proclaimed emperor some time before the death of

One wonders what Sidonius thought of the fall of his emperor and friend: in no letter preserved to us is there a word of regret or sorrow at its terrible suddenness. Nevertheless, it seems that the violent deaths which seemed the inevitable fate of emperors in these days had made Sidonius reflect somewhat on the doubtful chances of human fortune. Italy, as he said at a later time, had no luck in her emperors,[1] and when he received a letter from his friend Serranus, in which occurred a spirited laudation of the felicity of the emperor Maximus, he was moved to reply.[2] It was not only the death of Maximus, but those of his successors that had shown how foolish was such a doctrine of human felicity.[3] 'One has no right,' said Sidonius, 'to call those men fortunate who stand on the precipitous and slippery peak of the empire; one cannot say how many hours of misery are contained in the lives of the men who are called fortunate for this. If their state be the goal of happiness I know not, but it is certain that those who reach it are the most miserable of men.' Much of this letter is in the rhetorical tradition, yet there appears throughout a note of genuine feeling, which one would not have expected the death of Maximus in itself to evoke. Already Sidonius had passed beyond the cheerful optimism in which he had left for Rome in 455.

Majorian. That is not impossible, but *Fast. Vind. Priores*, 589 (i, p. 305) give Nov. 19th for the date, and they are a better authority.

[1] *C.* ii. 347–8 (Speech of Italy), 'quemcumque creauit / axe meo natum, confestim fregit in illo / imperii fortuna rotas'.

[2] *E.* ii. 13.

[3] That this letter was written after the death of Majorian seems fairly certain from § 3, 'antecedentium principum casus uel *secutorum*'. It would be possible, though barely probable, to date it after the death of Anthemius in 472; all the letters in Book ii, however, seem to precede Sidonius' election to the bishopric.

It would seem that Sidonius about this time resigned his office of *comes ciuitatis Aruernorum*[1] and perhaps his resignation is a symptom of the same disillusionment. If he had continued to hold an official position in Auvergne, the round of private visits which are a feature of his life during the next years (461–7) would hardly have been possible.

For in these years we see him no longer as a minor actor in the politics of the empire, but as a central figure in the social life of southern Gaul. We now enter into the full stream of his correspondence and instead of the official panegyrics we have the *vers de société*. We can now attempt to reconstruct his personal character not from his actions on the large stage of political life but from his own written words.

In this, however, we encounter a difficulty, which must be stated and faced by any inquirer into Sidonius' life and character. In the first letter of the whole collection, he tells us that his letters have received some kind of revision before publication both from his friend Constantius and from himself—'retractatis exemplaribus enucleatisque' are the exact words.[2] It is highly probable, as will be shown, that Book i of the letters was published before the rest, and of course the words in *E.* i. 1 only apply to the letters contained in it. We have proofs, however, that the later books have been similarly worked over. Thus in a letter to Lupus, bishop of Troyes, mentioning the collection of letters in Book vi, Sidonius points out that after they have been revised by Lupus they will no longer be regarded as his own,[3] and similarly it is stated that the letters in Book viii

---

[1] Cf. *E.* i. 11. 1, 'id iam agens otii'.
[2] *E.* i. 1. 1.   [3] *E.* ix. 11. 6.

have been revised by Petronius.[1] Finally, in the last
letter of the whole collection, that addressed to Genesius, he tells his correspondent that 'the incompatible
virtues of perfection and rapidity must not be expected:
for when a book is written, as it were, to order, the
author may perhaps expect credit for punctual delivery
but hardly for the quality of his work'.[2] This is not the
language that a man can use if he is merely turning
out his bureau for letters and publishing them just as
they are. Such a collection may be a book published
to order, but it is emphatically not a book written to
order. Thus we have evidence of a revision for Books i,
vi, vii, and ix; for the other books we have no such
explicit statements, nevertheless there is no reason for
supposing that they were not similarly revised and
every presumption that they were.

It is impossible to state definitely what is meant by
this revision; a passage from the letter to Lupus quoted
above hints that it involved, if nothing else, refinement
of grammar and diction. And there is another reason
for supposing that to some extent this was so. In the
whole of the nine books of the Letters there are no
fewer than 133 passages deriving from the letters of
Pliny and Symmachus,[3] and the majority of them are
rather direct quotations than mere verbal reminiscences. It is not easy to believe that a casual correspondence, even though many of the letters are very
formal, would contain so many quotations from two
prose authors. One may suspect that the revision

---

[1] *E.* viii. 16. 1, 'correctionis labor'.   [2] *E.* ix. 16. 3.

[3] See Gelzer, ap. Lütjohann, pp. 353–83. Many of the examples presented
by him are fanciful or may be explained as accidents. The number given in
the text represents those which I consider certain reminiscences. See below,
p. 171, n. 2.

involved in part an assimilation of the letters to those of Symmachus and Pliny, who were indeed Sidonius' professed models.[1]

It is possible that this assimilation has had an unfortunate effect on Sidonius as an historical authority; it may be that, in using a quotation from an earlier writer to describe a situation or a feeling, he has introduced a distortion from reality. Any large distortion of this kind is of course unthinkable, but small ones are quite imaginable. Whether this consideration has any validity, and, if so, to what extent, are questions that can only be subjectively answered, for we have no adequate contemporary source by which to control Sidonius' statements; but its existence must not be forgotten.[2]

Nor can we say definitely to what extent the revision involved changes of matter. It has been pointed out[3] that the letters are mostly divided up evenly among the correspondents, each man, as a rule, being only included as recipient of one letter: that being so, it is noteworthy that there are not the loose ends lying about that one would expect to find quite often, if the letters had been selected from a series and had been published without material revision. Each subject, in fact, as Sidonius says, 'ends with its containing letter'.[4] We may therefore fairly suppose that as a result of the revision each letter was made into a separate entity.[5]

[1] *E.* i. 1. 1; iv. 22. 2; ix. 1. 1.

[2] When such a borrowed passage is quoted in the text I shall note in brackets the reference to the author from whom it is taken.

[3] By Klotz in *P.-W.*, ii. A. 2, p. 2236.

[4] *E.* vii. 18. 4.

[5] Peter (p. 156) takes the view that when he revised the letters Sidonius omitted names which might give pain. He quotes *E.* ix. 6. 1 and *E.* ix. 7. 1. In the former passage he may be right, but in the latter he must be wrong. The general point—that Sidonius omitted proper names which were not important—may be true.

We can also notice in the letters a certain lack of intimate feeling:[1] many of them are more like moral essays than letters. It has been thought that they were conformed to the rules governing the art of epistolography as illustrated in the letters of Pliny and Symmachus,[2] and if that is so, it would not be only in language that Sidonius has followed them. There is very little, certainly, that is unconventional in the letters and practically nothing that is even momentarily startling; and it is not hard to believe that much of their spontaneity has been sacrificed to the conventions of an art form.

But whether the Sidonius which the letters present to us was different from the real Sidonius, we cannot definitely say. Nor, after all, from the historian's point of view does it very much matter. If Sidonius did touch up his letters to make his character more attractive, we can be certain that he must have done so conformably with the ideals of his age, and thus what the letters would lose in one respect they would gain in another. Even if doubts were cast upon the value of the autobiographical details presented in them, it could not be denied that they portray the life and ideals of the fifth century.

The letters of the period of retirement describe for the most part a pleasant and careless round of country-house life, divided between his ancestral property near Lyons[3] and the villa of Avitacum,[4] with numerous

---

[1] See Peter, p. 114 *et seq.* He notes that the description of the villa (*E.* ii. 2), the refusal to write history (*E.* iv. 22), and the preference for life in the country (*E.* viii. 8) are all stock motives. Cf. also *E.* iii. 13 and *E.* v. 7.

[2] See Klotz, l. c.

[3] Sidonius at Lyons, *E.* i. 5; ii. 8 (not certain but highly probable); ii. 10; iii. 12 (see Coville, pp. 35–6); iv. 25; v. 17; *C.* xii. He had a 'villula' in the country and a 'praedium suburbanum' (*E.* ii. 12. 2).

[4] Sidonius in Auvergne, *E.* i. 7; i. 11; ii. 2; ii. 6; ii. 9; ii. 14; iii. 12; *C.* xvii; xviii; xix; xxiv. On the site of Avitacum see Appendix B.

visits to the estates of his friends in south Gaul. Though
many of the letters are written from Lyons, and though
he speaks affectionately of 'Rhodanusiae nostrae',[1] he
was beginning to consider himself rather as an Arvernian. Thus to the Arvernian Ommatius he writes that
'thanks to your love for me, Christ has made this
Auvergne my own country',[2] and soon, when speaking
for the Arvernians, he slips quite naturally into the use
of the pronoun 'nos',[3] so completely had he identified
himself with his new country. Already he had shown
his admiration of the Arvernian men, when in the
panegyric to his father-in-law he had told how they
were invincible as cavalry or infantry;[4] already he
had described the fertility of the soil, to which even
Egypt, he said, was inferior;[5] and as he became better
acquainted with the land his praises of it remained the
same. It had, he said, 'a certain unique charm'; and
with perhaps a touch of personal feeling he describes
how a single glimpse of it would make a visitor forget his
own land.[6] We shall not well understand the uncompromising attitude which Sidonius took up in the face
of the Visigoths, unless we realize the depth of his
affection for the Arvernian soil.

But it is not only Auvergne and Lyons that he describes for us: the letters of these years, 461–7, show that
he was travelling much around central and southern
Gaul. There are mentions of visits paid to Bishop
Faustus at Riez, to the villas of his cousin Apollinaris
and Tonantius Ferreolus near Nîmes, to Narbonne,

---

[1] *E.* i. 5. 2; cf. *E.* i. 8. 1.

[2] *C.* xvii. 19–20, 'dabit omnia Christus, / hic mihi qui patriam fecit amore
tuo'. On the translation and interpretation see Coville (p. 37, notes 1 and 2).

[3] *E.* iv. 21. 3; cf. *E.* ii. 2. 1, 'regionis nostrae'.

[4] *C.* vii. 149–50.   [5] Ib. 141–6.   [6] *E.* iv. 21. 5.

and still farther afield to Toulouse, Bordeaux, and Bourg-sur-Gironde in the Visigothic kingdom.[1]

These visits are described in the poems and letters, and doubtless there are many more that are unrecorded. Sidonius had friends all over Gaul, and if he thought a man's friendship worth making, he spared no pains to get it.[2] That there were plenty of houses in which he could expect entertainment is seen from the dedication verses which accompanied the publication of a section of his poems.[3] He describes how the book sets out for his friends at Narbonne: it goes a circuitous way through Javols and past Rodez, and it stops at no fewer than twelve places on the journey. Where the book was welcome, we may be sure that the man was welcome too.

If we attempt, however, to fit these visits into a chronological scheme, we meet with difficulties that are often insurmountable. When he was preparing his letters for publication, Sidonius brought them out of his writing cases in quite a haphazard way.[4] Occasionally two or three letters are placed together which are obviously connected with each other,[5] but as a general rule, it seems that in their arrangement chronological succession is quite disregarded.[6]

[1] The evidence for these movements is derived from scattered allusions in the letters and poems. Riez, *C.* xvi; Nîmes, *E.* ii. 9; Narbonne, *C.* xxii and xxiii; Toulouse, *E.* iv. 24; Bordeaux, *E.* viii. 11 and 12; Bourg-sur-Gironde (Burgus Pontii Leontii, see Naufroy, *Histoire de Bourg-sur-Gironde* (1898), p. 9), *C.* xxii. The passages cited by Dalton (i, p. lxxxiii, n. 1) are inadequate to prove a visit to Lérins (*E.* ix. 3. 4; and *C.* xvi. 105 et seq.). Sidonius had friends in Spain (*E.* viii. 5, and perhaps *E.* ix. 12), but there is no evidence that he visited it.

[2] Cf. *E.* v. 11. 1 and *E.* vii. 14. 9.

[3] *C.* xxiv. Compare the list of Sidonius' friends living near Narbo given in *C.* xxiii. 443–86.

[4] Cf. *E.* viii. 1. 1; *E.* viii. 16. 3; *E.* ix. 1. 4; *E.* ix. 16. 2, and see below p. 171.

[5] Cf. *E.* i. 3 and 4, 5–11; iv. 2 and 3; iv. 4 and 6; v. 6 and 7; viii. 9–12, &c.

[6] Attempts at dating the letters have been made by Baret (pp. 125–45),

Such chronological details as the letters and poems themselves give are few. That Sidonius was at Lyons fairly soon after the banquet of Majorian seems shown from the verses at the end of *C.* xii.[1] In this poem, which is addressed to his old Arvernian friend Catullinus, Sidonius gibes at the seven-foot high Burgundians who smear butter on their hair, and at the end of it he says 'My Muse must now be silent or some one will be calling these verses satire too.' Catullinus, it will be remembered, was the man responsible for the affair at Arles in 461, and this fact, combined with the language of the concluding words of the poem, makes it obvious that Sidonius has the incident of the satire very close in his mind, and, what is more, he expects that the recipient will see the point too.[2] The journey to Bordeaux mentioned in *E.* viii. 11 and 12 can also be dated reasonably closely. Sidonius asks his correspondent Trygetius whether he has so soon forgotten his recent journey to Calpe and the south of Spain,[3] and the words 'fixa tentoria in occiduis finibus Gaditanorum' suggest some kind of military expedition. The only campaigns in Baetica of which we have record about this time are in 458 and 459,[4] and as our information about Spanish wars in this period is very detailed, we may infer that there were no others. The campaign of Trygetius will

Chaix, and Dalton, but their three systems are alike in this alone, that they date many letters far too precisely.

[1] *C.* xii. 20–2, 'Sed iam Musa tacet tenetque habenas / paucis hendecasyllabis iocata, / ne quisquam satiram uel hos uocaret'. Sirmond noticed the allusion (in ed., p. 144) and is followed by Hodgkin (ii, p. 363, n. 1), but by no one else. Some date the poem to 470–2 (Chaix, i, pp. 315–16; Baret, p. 154); Fertig (ii, p. 17) sees a reference to the Burgundian garrison of Clermont in 474, and Coville looks back to 458 (p. 127).

[2] That Sidonius had no objection to writing satire in the ordinary way is seen from *E.* iv. 18. 6; cf. *E.* v. 8. 1.

[3] *E.* viii. 12. 2.   [4] Hydatius, 192, 193 (ii, p. 31).

then have occurred in one of these years. 'Celeriter' and 'Nuper' in Sidonius' letter suggest a date not much, if at all, later than 461–2 for this visit, which is placed in the winter;[1] and as Pontius Leontius seems to have been Sidonius' host[2] at the time when the letter was written, it is likely that the visit to Pontius' home at Bourg-sur-Gironde (C. xxii) occurred in the same year. A rough *terminus ante quem* for the Bourg poem is provided by the statement[3] that Narbo, formerly called 'Martius', has lately become entitled to the name. This can only refer to the capture of Narbo by Theodoric in 462,[4] and thus C. xxii will date from shortly after that. We can hardly suppose a great interval of time between C. xxii and the visit to Bourg which it describes, and thus we get corroboration for our theory that this visit is to be dated to winter 461–2. C. xxii again serves to give us a rough date for the visit to Narbo (463?) and from this we can date C. xxiii. Our chronology, based indeed on such vague words as 'nuper' and admittedly conjectural, may be worked out in this way:

461 (spring). Arles with Majorian.

461 (summer). Lyons (C. xii).

? 461–2 (winter). Bourg (E. viii. 12).

463. Narbonne (C. xxii and xxiii).

Whether the visit to Toulouse took place during 461–2 or whether Sidonius made another journey into Visigothic territory we cannot say:[5] no more can we assign even approximate dates for the visits to Riez and Nîmes.

[1] 'temporibus hibernis', E. viii. 12. 1.   [2] Ib. 5.

[3] '... Narbonem quondam Martium dictum, sed nuper factum', C. xxii, ep. 1.

[4] Hydatius, 217 (ii, p. 33). For the date see Tammasia, p. 21.

[5] E. iv. 24. Dalton (ii, p. 42), Chaix (ii, pp. 234–7), and Baret (p. 135) all differ as to the date of this letter. In truth even Chaix's date, 476–8, though improbable, is not out of the question.

Nevertheless, the accurate determination of the chronology, even if we could do it, would be a jejune achievement: more interesting by far is to attempt from the numerous details which Sidonius gives us to reconstruct the social life of these great landed proprietors.[1]

They were early risers: the barbarian king, Theodoric, was usually astir before day-break[2] and after early prayer was ready for administrative work, but he was a busy man. Still Sidonius does not consider it unusual that a public function or a church festival should start at sunrise or even before.[3] The wealthy Consentius, who had no official business, was up in the early morning, and had already started playing games by the fourth hour;[4] and the guests at Avitacum, rather surprisingly, as we may think, were told how delightful it was to hear the rooks salute the dawn.[5]

Games in the morning seem to have been the common practice. At Vorocingus and Prusianum, the villas of Tonantius Ferreolus and Sidonius' cousin Apollinaris, the morning visitor might arrive to see four of the villa's occupants playing a game like tennis on the lawn.[6] After the festival of St. Justus at Lyons, the younger members of the party settled down to a game of 'harpaston',[7] a form of exercise which demanded, says Sidonius, that the participants should be in good training. Too much of it for an old man was not good for the liver. The older men, as a rule, would amuse themselves with the dice-box,[8] and there was a game

---

[1] See Fertig, pp. 21–4; and Dalton, i, pp. xcv–cv.
[2] *E.* i. 2. 4.   [3] *E.* viii. 6. 5; v. 17. 3.
[4] *C.* xxiii. 487.   [5] *E.* ii. 2. 14.   [6] *E.* ii. 9. 4; cf. *E.* ii. 2. 15.
[7] *E.* v. 17. 6. On the game and its rules see Marquardt, *Privatleben*, p. 846.
[8] *E.* ii. 9. 4; v. 17. 6; *C.* xxiii. 491. At Ravenna, where everything according to Sidonius was abnormal, the old men played at ball, the young at dice, *E.* i. 8. 2.

played with dice and a board, somewhat like back-gammon, at which Theodoric had learnt to be very proficient.[1] The Gallic noble was passionately fond of gambling, and though Lampridus, the poet of Bordeaux, professed himself bored by it,[2] another Aquitanian friend, Trygetius, actually expected to see a dice-box, board, and men in the boat that was to take him to his host's mansion.[3]

The mid-day dinner (*pransum*) occurred at about eleven o'clock.[4] It was still eaten in the old Roman manner, with the guests reclining at table,[5] and it was the fashion in aristocratic circles to serve numerous courses on small dishes;[6] this gave opportunities for displaying the silver plate in which great families took pride.[7] Still there was no doubt that there was enough to eat: the full Gallic table was proverbial.[8] At Vorocingus there were boiled and roast meats.[9] Fish was very popular fare: Trygetius, the gourmet, is invited to partake of the mullets and crabs of the Garonne, but to make his visit more pleasant he is urged to bring with him the greater delicacies of his own district, the lobsters of Bayonne, and in addition oysters were to be brought specially for him from Bordeaux.[10] Avitacum was regularly supplied with trout from the lake,[11] and at his home in the Lyonnais Sidonius might expect gifts of fish from his brother-in-law Agricola.[12]

[1] E. i. 2. 7; cf. 'tabula', E. v. 17. 6; 'tesserae', E. ii. 9. 4.
[2] E. viii. 11. 8.   [3] E. viii. 12. 5.
[4] E. ii. 9. 6. The time was measured by a clepsydra.
[5] E. i. 11. 10.   [6] E. ii. 9. 6, 'senatorium ad modum'.
[7] E. i. 2. 6; ii. 2. 11; viii. 7. 1; ix. 13. 5, vv. 54–7; C. xxii, ep. 5.
[8] 'abundantiam Gallicanam', E. i. 2. 6. On the voracity of the Gauls, see Amm. Marc., xxii. 12. 6; Sulp. Severus, *Dialog.*, i. 8. 5, 'edacitas in Gallis . . . natura'. Cf. ib. 4. 6; 9. 2.
[9] E. ii. 9. 6, 'edulia nunc assa nunc iurulenta'.   [10] E. viii. 12. 1, 7.
[11] E. ii. 2. 12; cf. C. xviii. 10; C. xxi.   [12] E. ii. 12. 1; cf. E. ii. 9. 9.

. In general the food, compared with the proverbial luxury of Byzantium and the East, was regarded as simple.[1] It is noteworthy indeed that there is very little mention of imported delicacies in the writings of Sidonius. In a poem addressed to his friend Ommatius he says that in his house the visitor must not expect African bread[2] or the wines of Gaza, Chios, and Italy:[3] unless this is merely conventional phrase, it implies that there were houses in Gaul at which these delicacies could be obtained. The only imported food which is definitely mentioned in the letters is the humble salt which seems to have been brought from the Tarragona mines.[4]

Men who had taken up with the new fashion of asceticism for the most part avoided meat: Vectius who was, as Sidonius says, 'a monk in all but his clothes', would not eat game,[5] and Maximus, newly ordained a priest, refused anything but a vegetable diet.[6] Both of them, however, were tolerant enough to provide meat courses for their guests. From the clergy a higher standard of abstinence was naturally expected: the Abbot Chariobaudus, to whom Sidonius addressed a letter, is said to be exhausted by long fastings,[7] and Sidonius himself, when bishop, agreed that clerics should fast on alternate days.[8] Still, even monks did not usually practise the extreme abstinence of the East, for we may remember that the strict rules of asceticism were modified by St. Benedict for the Western monasteries.[9]

---

[1] *E.* iv. 7. 2.

[2] On imports of foodstuffs from Africa at an earlier date see *Const. Honorii a.d. 418* in Haenel, *Corpus Legum*, p. 238.　　　　[3] *C.* xvii. 13–16.

[4] *E.* ix. 12. 1.　　　　[5] *E.* iv. 9. 3.　　　　[6] *E.* iv. 24. 3.

[7] *E.* vii. 16. 2.　　　　[8] *E.* vii. 14. 12; cf. Ruricius, *Ep.* i. 9. 2.

[9] See Butler in *C.M.H.*, i, p. 536.

The lower classes seem to have eaten chiefly vegetables, but from compulsion. In one of his rare references to them, Sidonius speaks of their meal of onions.[1]

Sidonius does not give us much information about the drinks taken at table, and it is a surprise indeed that, when mentioning the delights of a table in the Médoc, he says not a word about the wine.[2] There are several references, however, in other passages to vineyards,[3] and Pontius Paulinus kept a cellar that Sidonius in a poem urges Bacchus to visit.[4] In a passage from the poem to Ommatius quoted above imported wines are mentioned as the great delicacies[5] and at the supper party of 460 Falernian alone is mentioned.[6] There is reference in a poem, however, to a district near Lyons, from which came celebrated wines of ancient repute.[7] Sidonius compares some of his work which has been hastily done to immature wine,[8] and from this we should doubtless conclude that wine was usually allowed to keep. The drinking of beer, which struck the notice of the emperor Julian,[9] is not mentioned in Sidonius. Probably its use, in the south at least, was confined to the lower classes.[10] The guests at Avitacum, to our, and perhaps their own surprise, were expected to drink water.[11]

After dinner followed the siesta, the name indicates the hour; Theodoric did not rest long, we are told, but he was not at work again until the ninth hour

---

[1] *E.* iv. 7. 2. It is curious that onions (with leeks) are mentioned as the food of barbarians (*C.* xii. 14). They also ate sausages, *E.* viii. 11. 3, ver. 46.

[2] *E.* viii. 12.   [3] *E.* i. 6. 4; ii. 9. 1; v. 20. 4; viii. 4. 1.

[4] *C.* xxii. 218-20.   [5] *C.* xvii. 15-16.   [6] *E.* ix. 13. 5, ver. 59.

[7] *C.* xvii. 17-18.   [8] *E.* viii. 3. 2.

[9] Julian in *Anth. Pal.*, ix. 368 (Bidez and Cumont, 168).

[10] On beer (*caelia*) prepared for the farm-labourers by an Arvernian master, see Greg. Tur., *de Gloria Conf.*, 1.   [11] *E.* ii. 2. 12.

(three o'clock).[1] In the afternoon, the host and his guests might go out riding, and perhaps hunt, either with hound or hawk, if they had not decided to make a full day's hunting-expedition. In spite of his austere habits, Vectius was renowned as a judge of hounds or a trainer of hawks.[2] The emperor Avitus and Ecdicius, as we have seen, were famous for their exploits in the chase. The noblest game was the boar, which was hunted with the spear. Of less account were stag-hunting and hare-hunting.[3]

After their exercise, the men would return ready for the bath before supper; and if they were feeling lazy they might spend an hour or more conversing in the bath-rooms.[4]

Supper (*cena*) followed: in some houses it was the custom to enliven it by discussing Appuleius' work, *Problems for Banquets*;[5] in other houses there were performances on a hydraulic organ[6] or exhibitions of dancing and singing.[7] If there was company, one or more of the guests might be called upon to deliver a poem.[8] Soon after supper all went to bed, and the day was over.[9]

[1] *E.* i. 2. 7, 9.   [2] *E.* iv. 9. 2.   [3] *E.* viii. 6. 11, 12.

[4] *E.* ii. 9. 9, 'hic nobis trahebantur *horae*'. On baths see *E.* ii. 2. 4–9; *C.* xxii. 184–6; *C.* xxiii. 495–9. The bath at Avitacum had four rooms—*frigidarium, unguentarium, caldarium,* and *baptisterium*. This type of bath is quite common throughout the empire (see Cagnat-Chapot, pp. 210, 212). According to Sidonius his *frigidarium* was as good as that of a public bath (*E.* ii. 2. 5 = Plin., *Ep.* ii. 17. 16). When baths were out of order, a makeshift could be constructed by filling a pit with hot stones and laying skins over it (*E.* ii. 9. 8 with Dalton's good note, ii, p. 225).

[5] *E.* ix. 13. 3, 'formulas conuiuialium quaestionum'. Cf. Macrob., *Sat.,* vii. 3. 24.

[6] *E.* i. 2. 9. On these 'organa hydraulica' see Darembourg et Saglio, iii, p. 212.   [7] *E.* i. 2. 9; ix. 13. 5, vv. 63–81.

[8] *E.* ix. 13. 4; cf. *E.* i. 5. 3, 'Phaethontiadas cantatas saepe *comissaliter*'.

[9] Sidonius nowhere mentions anything done after supper and Theodoric certainly went to bed at once (*E.* i. 2. 10).

Such was the ordinary life of the fifth-century Gallic aristocrat, as Sidonius depicts it. One feels that with the exception of its customary drunkenness[1] he might be describing the day of an eighteenth-century squire. And not a few of these Gallic land-owners too had literary tastes. Consentius at Octavianum had a fine library in which he worked hard writing poetry,[2] and Polemius, praetorian praefect of the Gauls, would employ his leisure in the study of philosophy.[3] Sidonius has warm praise for the library of Tonantius Ferreolus at Prusianum, in which the Roman classics and the more modern devotional works were on opposite sides of the room,[4] and doubtless there were others like him.[5] A devout aristocrat might peruse the scriptures, even while eating his food.[6]

Of the life led by the women of this period we know tantalizingly little. To speak of a woman's life, says Sidonius,[7] demands both delicacy and reticence, and he has kept his word to us. The pessimistic Salvian[8] might declare that the Aquitanians all treated their wives worse than their slaves, but it does not appear that the aristocratic woman in Sidonius' day was a mere drudge. It is true that she was expected to give much of her time to the loom;[9] still, she is allowed by Sidonius as an ideal that same 'libertas grauis et pudor facetus'[10] which he attributes to the discourses of his male friends[11]—and not unnaturally, for, though we are ignorant of the

---

[1] The drunkenness of the Gauls was well known to ancient historians (Diod. Sic., v. 25; Pliny, *N.H.*, xiv. 22; Amm. Marc., xv. 12. 4), but there is little trace of it in the letters (see *E.* ii. 9. 8; iii. 13. 4).

[2] *E.* viii. 4. 1. [3] *E.* iv. 14. 2; cf. *E.* i. 6. 5; iii. 6. 2.

[4] *E.* ii. 9. 4–5. [5] Cf. *C.* xxiii. 263–306. [6] *E.* iv. 9. 3.

[7] *E.* vii. 9. 24. [8] *De Gub. Dei*, vii. iii. 16.

[9] *C.* xv. 126–78; xxii. 194–203. [10] *E.* ii. 8. 3, vers. 11.

[11] *C.* xxiii. 438.

facts in Sidonius' own day, we know that Ausonius'
mother had been educated (and flogged) with the boys
in the grammar school.[1] Eulalia, who is compared to
'Cecropian Minerua',[2] may be exceptional; but Hes-
perius is told that marriage will not hinder but help his
studies,[3] and Tonantius Ferreolus' wife, Papianilla,
was a partner who shared his troubles.[4] In all his visits,
it is interesting to notice, Sidonius never seems to take
his Papianilla with him; still a wife was sometimes included in an invitation,[5] and women accompanied
their menfolk on pilgrimages.[6] The cult of asceticism
had extended to women, and the virgin had a chance
of displaying her individuality which was hardly possessed by the unmarried woman of earlier days.[7]

The salient fact of this country-house life is the extraordinary feeling of security that seems to accompany it.
We could scarcely believe that the fathers of the men
who were Sidonius' hosts had lived through the terrible
years from 400 to 450. And there is another even
stranger fact about these noblemen: it is no accident
that the picture given of Theodoric the Visigothic king
fits in so completely with that of the Gallo-Roman
nobles: Trygetius of Bordeaux, Consentius of Narbonne
were actually living in the territory of a barbarian
king.[8] Yet we could hardly realize this, if it was not
confirmed by external evidence.[9]

[1] Ausonius, xiii. 2 (*Protrepticus*). 33.   [2] *C.* xxiv. 95–6.
[3] *E.* ii. 10. 6.   [4] *C.* xxiv. 37.   [5] *C.* xx. 3; *E.* iv. 18. 2.
[6] *E.* iv. 6. 2.
[7] Devout women: Frontina, *E.* iv. 21. 4; Eutropia, *E.* vi. 2. 1; Faustus'
mother, *C.* xvi. 84–90.   [8] This point is well put by Dalton, i, p. lxv.
[9] A remarkable fact in the letters of Sidonius is the almost complete absence
of any reference to barbarian 'hospites' (*E.* ii. 1. 3, where 'hospitibus' is the
correct reading, is the solitary mention, though the 'patronus' of *C.* xii. 11 may
be, as Fustel de Coulanges thinks, a 'hospes'). It may be the fact that the numbers of the Gothic population were inadequate to permit the settlement of

One of the clearest indications of the security of a civilization is the state of its communications; and judged by this test the Gallic civilization of the fifth century comes out fairly well. There was still the old distinction between the imperial roads[1] and the lesser ways of the country-side:[2] Sidonius tells us, however, that the imperial columns beside the former were sufficiently old and that the emperor's name on them had become green.[3] Such milestones with the imperial inscription were put up when a road was repaired[4], and Sidonius' remarks are probably a reproach on the length of time since the last repair was made: when the priest Constantius made his journey from Lyons to Clermont, the road was full of holes.[5] On the other hand, if the emperor did not repair the roads, the officials in the departments of the local governors did what they could. Seronatus, the rascally governor of Aquitanica, is seen sending an official on in advance to make the contractor hurry with the repair of a road on which he wished to travel.[6] To Sidonius, who hated the man, this does not count for righteousness;[7] nevertheless we cannot doubt that what a bad governor would do for his own convenience a good one would do for the public benefit. The imperial post service was

'hospites' in the newly acquired territories, such as *Narbonensis prima*, where Consentius lived (so Schmidt, i, p. 190). Still we may be fairly certain that on the lands of Trygetius and Pontius Paulinus a Gothic 'hospes' was residing.

[1] As was the fact down to the era of railways, journeys, where possible, were made by water (*E.* i. 5. 3; ii. 10. 4, vv. 22, 25-7; vi. 8. 1; vii. 10. 1; viii. 11. 3, ver. 31).

[2] *C.* xxiv. 5-9; cf. *E.* ii. 9. 2.

[3] 'satis uetustis / nomen Caesareum uiret columnis', *C.* xxiv. 6-7.

[4] See Cagnat-Chapot, i, pp. 50-2.

[5] *E.* iii. 2. 3. A certain allowance must be made for rhetorical exaggeration in this letter.

[6] *E.* v. 13. 1, 'Euanthius . . . iam contractas operas cogit eruderare'.

[7] Ib. 4, 'mihi latronis et beneficia suspecta sunt'.

still in existence, for Sidonius, as we shall see, used it on
his journey from Lyons to Rome in 467;[1] indeed, it
survived the empire and was still maintained by the
Visigothic kings.[2] The 'mansiones', or official rest-
houses, were still used in 473 on the Lyons-Clermont
road.[3] Sidonius' slaves, who were encumbered by
tents and heavy baggage, made eighteen miles between
dawn and lunch-time (fifth hour)[4] and that is very fair
going for unsprung carts. From these facts one should
draw the inference that, while the road system was less
good than it had been in the best days of the empire,
it was still in quite fair condition. The information
given us by Sidonius only applies, however, to the
centre and south of Gaul: in the north, the roads had
been bad a hundred years before, and they were
probably very bad now.[5]

Not only were the roads, at least in the centre and
south, in fair condition, but the travelling seems to have
been tolerably secure. There is, it is true, one mention
of brigandage in the letters,[6] but Sidonius nowhere
hints, except in time of war, that travelling was
likely to be dangerous either to himself or to any of his

---

[1] *E.* i. 5. 2.

[2] *Leg. Vis.*, v. 4. 19; *Lex Rom. Vis.*, viii. 2. See Dahn, *Könige der Germanen*, vi.
pp. 285–6, and in general Hudemann, *Das römische Postwesen*, pp. 46–51.

[3] *E.* iii. 2. 3; cf. *E.* v. 7. 3 ('veraedarii').

[4] *E.* iv. 8. 2. It is very unfortunate that owing to the somewhat woolly
verbiage of Sidonius it is not easy to decide whether the eighteen miles represents
the distance travelled before lunch or in the whole day. I think that the former
is correct—for this reason: Sidonius says that his servants had gone ahead and
pitched their tents eighteen miles away, that he followed after and feared that
he would be late for his lunch ('sero pransuri'). Now a man who is carrying
his own lunch can eat when he likes, so Sidonius can only mean that he feared
that he would be late in reaching the camp of his servants for lunch.

[5] *Pan. Lat.*, v. 7. 1–4 (ed. A. Baehrens, 1911); cf. 'difficultas itineris' in a
letter to Principius, bishop of Soissons (*E.* viii. 14. 8).

[6] The mysterious 'Vargi' of *E.* vi. 4. 1.

friends. At a time when there were Gothic raiders attacking Auvergne, a traveller came through Clermont on his way from Riez to Brittany (or possibly even to Britain).[1] Even more startling, perhaps, is the fact that a journey to properties near Bayeux is not considered as calling for any kind of special comment: yet we can hardly doubt that the whole country around Bayeux was infested with Saxons.[2]

And there is another fact emerging from the correspondence which argues such an ease of communication as one would hardly have expected for this period. It is a fact that pervades the whole correspondence, so large, indeed, that it is apt to escape notice.

We possess 157 letters of Sidonius: in practically none of them does he assume that their carrier will fail to reach his destination.[3] Letters are sent far and wide, sometimes to the north, to Rheims,[4] or even to Trier,[5] but it is never hinted that there will be any difficulty in the transmission either of the letter or of the reply. It may be stated that correspondence with a distant friend is difficult because there are few travellers going that

[1] E. ix. 9. 6. If this were so, it would be the only reference in Sidonius to communication with our island, but it cannot be accepted as at all certain. There were certainly Britons in Armorica at this time (see Loth, L'Émigration bretonne en Armorique, p. 50), and a man travelling from Riez through Clermont is more likely to be making for Armorica than for Britain. On the other hand, it has been pointed out that the traveller Riochatus may be identical with St. Riochatus of early Welsh hagiology (Tillemont, Mém. Ecclés., xvi, p. 421), and when Sidonius tells Faustus that Riochatus is going 'Britannis tuis', it suggests the island, for there is some evidence that Faustus was a Briton from the island (see Vita S. Germani, ap. Nennium, ii, 48 in Chron. Min., iii, p. 192; cf. Duchesne in Revue Celtique, xv, p. 187). The question must unfortunately be left open.

[2] E. iv. 18. 2. On the 'Saxones Baiocassini' see Greg. Tur., Hist. Franc., v. 19. (26); x. 9; cf. Longnon, pp. 172–5.

[3] The only exception, and that perhaps is not a real exception, is the statement made to Principius that 'difficultas itineris' (see above, p. 76, n. 5) may hinder correspondence.   [4] E. ix. 7.   [5] E. iv. 17.

way,[1] or that one cannot be particular in the choice of
letter-carriers;[2] a messenger may lose a letter by his
own negligence,[3] or in very exceptional circumstances
he may be stopped on the way and his bag examined
by a Visigothic or Burgundian official.[4] Only when
Clermont was actually besieged by the Visigoths did
Sidonius fear for the safety of his letters[5] and even then
he could communicate with his friends, and when an
answer failed to come he was more inclined to blame
the taciturnity of his correspondent than the insecurity
of travel.[6]

It was an easy life that these Gallo-Roman aristo-
crats led: in fact in the very ease of it lay its chief danger. With the duties of fighting given over to the barbarians, and much of the local administration either in
the hands of the middle-class *curiales* or of the bishop,
there was a great temptation to idleness. Many, as has
been said, amused themselves with writing poetry or
studying philosophy, but both of these must have
palled, when the subjects were so artificial. A few
amused themselves with building,[7] that ancient hobby
of the idle rich, but one cannot always build. Visitors
came and went, it is true, but visitors are an uncertain
stand-by for intellectual food. Many squared their
shoulders and bent down to hard agricultural labour
on their estates[8] and perhaps they were the best of all.
Still, such a life is hardly conducive to intellectual
vigour, and there was truth in Sidonius' rebuke that
such a man was not the master but the slave of his land.[9]
It was no accident that the men of greatest intelligence

---

[1] *E.* ii. 11. 1. [2] *E.* iv. 7. 3. [3] *E.* iv. 12. 2.
[4] *E.* ix. 3. 2 and ix. 5. 1. [5] *E.* vi. 6. 1.
[6] *E.* iii. 7. 1; iv. 5. 1. [7] *E.* iv. 15. 1; v. 11. 2; viii. 6. 10.
[8] *E.* i. 6. 4; ii. 14. 2; vii. 15. 2; viii. 8. [9] *E.* viii. 8. 2.

among the Gallo-Roman nobles are found entering the
Church. The Church, at least, gave them more scope
for thought and action.[1]

It is a notable characteristic of the upper-class life in
fifth-century Gaul that it centres around the country-
house. It is easy in reading through the letters of
Sidonius to see that he is describing a society in which
town life is of secondary importance. Many nobles, it
is true, including Sidonius himself[2] had town houses;
but with the exception of Arles, the seat of the prefecture, we hear little of the town except as a centre of
ecclesiastical or official life.[3] Sidonius reproaches one
of his correspondents with spending most of his time
in the country,[4] but the only visit which he himself is
recorded as making to Clermont, before he became
bishop, was made on the occasion of 'an unpromising
piece of business'.[5] And though archaeologists may
have put the point too strongly, there is no doubt that
the fourth and fifth centuries were an era of decline for
the towns.[6]

One of the effects of this was to deprive the large
landowner of opportunity for the exchange of intellectual thought; and thus the culture of the great houses
was not strengthened by the introduction of new ideas.
Being for the most part pagan and artificial, it was
not based on any deep spiritual impulse, and on that

---

[1] As Arbogastes (*Gallia Christiana*, ii, p. 481), Maximus (*E.* iv. 24), Ruricius
(Krusch in *M.G.H.*, pp. lxii–lxiii), and of course Sidonius himself. Cf. *E.* vii.
12. 4.

[2] *E.* ii. 12. 2; iii. 3. 5; vii. 15. 1.

[3] Like most generalizations this must not be pressed too hard: in *E.* v. 17. 1
Sidonius mentions the delights of the town: 'te auocat uenatio *ciuitas* ager', but
such a passage is exceptional.   [4] *E.* viii. 8. 1.   [5] *E.* ii. 14. 2.

[6] See Blanchet, *Les Enceintes romaines de la Gaule*, p. 300 et seq.; C. Jullian,
Histoire de la Gaule, vii, p. 14.

account its life must always have been precarious. Moreover, it was exotic. As one glances through the poetry of Sidonius, it is remarkable how many of his similes are taken from Greece and the East: this oriental colouring is one of the most curious features of his work. The poem to Consentius[1] is particularly instructive in this respect, and an enumeration will best show the validity of an argument, which is obvious to a student of Sidonius' works, but not easy to make convincing to a casual reader. In the 512 lines there are no less than 63 references to Greek localities or Greek men; that is, one in every eight lines. And this is in a poem addressed to a Gallic poet who had asked Sidonius to write something in his honour. *C.* xxiv, the last in the collection, tells the same tale; in a poem describing the reception of the collected poetry in various Gallic mansions there are twelve references to Greece or the East in 101 lines.[2] We should hardly expect the Biblical quotations in a patristic treatise to be much more numerous. And the Bible was a real inspiration to the fathers, but we can hardly believe that the Greek mythology touched the hearts of Sidonius and his friends with more than a very faint glow.[3]

Moreover, this essentially artificial culture was founded on a very narrow base. With the decay of the middle classes, which had been proceeding throughout the past two centuries,[4] there had disappeared that society of educated townsmen which might have given even such an artificial culture stability. In a passage,

---

[1] *C.* xxiii.

[2] Some of them, such as the reference to Porus in vv. 72–4, are introduced with only the barest relevance.

[3] Note as contrast that there is no mention of the Gallic hero Vercingetorix in the works of Sidonius.   [4] See Dill, pp. 245–81.

the soundness of which, it is true, may have been affected by borrowing, Sidonius complains how the town councillors of Clermont are for the most part men of exceeding illiteracy,[1] and though this may be exaggerated, it can hardly be quite untrue. And below them there were only the brutish *coloni*.[2]

A culture which was confined to such a small class of persons, and derived its inspiration so largely from sources distant from it in time and space, must have been endangered by the lightest shocks, and the feeling of its unreality must have increased as communication with Greece and the eastern lands became ever more difficult.[3] Even in Sidonius' own time, even in the years when, as we have shown, life was comparatively secure and internal communications easy, we can paint from the letters a picture of continual cultural decay. In the northern regions, where intercourse was harder and the barbarian infiltration stronger,[4] the decay was more pronounced. In a letter written at a later date, about 470, to Arbogastes, Count of Trier,[5] Sidonius says bluntly that the Roman tongue has long been banished in Belgium and on the Rhine, and that Arbogastes alone preserves what remains of Roman culture and even of Roman speech in the north.[6] In the south, where the barbarians were less widely distributed, the state of affairs was less gloomy.

[1] 'turba numerosior illitteratissimis litteris uacant', *E*. iv. 3. 10 (= Pliny, *Ep*. i. 10. 9).

[2] On the lack of culture among the lower classes see *E*. iv. 7. 2; iv. 17. 2.

[3] On the difficulties of communication see Baynes in *J.R.S.*, xviii (1928), p. 225. A contrary view is taken by Pirenne (*Les Villes du Moyen Âge*, chap. i). But see Baynes's reply, *History*, xiv, no. 56, pp. 290–8.

[4] See Kurth, *Études franques*, iii–xiii.

[5] See *Epist. Auspicii, ap. M.G.H. (Ep. Mer. et Kar. Aeui*, i), pp. 135–7; cf. Sirmond, p. 49.

[6] *E*. iv. 17. 2, 'uanescentium litterarum' (= Pliny, *Ep*. viii. 12. 1).

But throughout the country literature was losing ground. When encouraging a young orator, Sidonius asks himself almost against his will whether there will be any more students to hand on the torch of Latin erudition, 'for there are few men now,' he says, 'that have respect for culture'.[1] 'The numbers of the indifferent,' he says in another passage,[2] 'are growing so great that soon the Latin tongue will be entirely dead.' Ecdicius is congratulated for reintroducing a purer Latin among the nobles of Auvergne.[3] One may suspect that the exaggerated praises which Sidonius gives to contemporary writers may have been the desperate encouragements of a man who knew well that he was leading a forlorn hope, and when he says that to diligence in literature he always awards the highest praise of which he is capable[4] he practically admits the fact. Indeed, when we compare the polished Latin of Sidonius with the wild illiteracies of the Arvernian Formulae,[5] the earliest of which was composed in the generation after his, we can only admit that his fears were justified. The accuracy, the comeliness, and the grandeur of the Latin tongue,[6] that he had laboured so hard to defend, was disappearing. His age, he said, was barren, and the bones of his generation lacked the marrow of their ancestors; the world was old.[7] The conservative may be forgiven a regret for a culture that was passing away, and we may pardon Sidonius' very

---

[1] E. v. 10. 4.   [2] E. ii. 10. 1; cf. E. ix. 11. 7; viii. 2. 1.

[3] E. iii. 3. 2, 'Sermonis Celtici squamam'. On the question of the survival of the Celtic language after the Roman conquest see Jullian, *Hist. de la Gaule*, vi, pp. 110–15.

[4] E. ii. 10. 1; cf. Ruricius, *Ep.* i. 8. 6; i. 9. 1.

[5] See *M.G.H., Leges*, v, pp. 28–31 (ed. Zeumer).

[6] 'Scientia pompa proprietas linguae Latinae', E. iii. 14. 2. Cf. E. iv. 17. 2.

[7] E. viii. 6. 3 (= Pliny, *Ep.* viii. 12. 1).

real injustice to the Church of which he was a member
when he wrote these words. But applied to the old
pagan culture his words were true. It had died of old
age, and when we remember that characteristic of
the old is said to be their self-centredness and lack of
sympathy with present conditions, we may perhaps
think that Sidonius spoke truer than he knew.

Now that we have used his works to construct a picture of the social life in which he moved, and the culture on which his intellect was nourished, it is time to
examine the character of the man himself. 'A man's
book,' says Sidonius,[1] 'reflects his mind, as a mirror his
face.' Let us then take him at his word and see what
kind of man results from a study of the letters and
poems. The real Sidonius may elude us, but we should
be able to capture an image of the man, as he wished
posterity to see him. In one of his letters Sidonius has
given to a friend just such a picture of himself.[2] It is
the picture of a likable but not very deep character.
A hospitable and cheerful man, kindly to his house-
hold, and delighting in the society of learned men—
such is what he wishes us to think him, and the picture
is borne out on the whole by the rest of the correspondence. He enjoyed his round of visits, and even after he
had become bishop he made no secret of his love for
good cheer;[3] he must have been good company, if, as he
tells us, landowners would picket the roads on which
he was travelling to secure him as a guest.[4] But if he
made use of the advantages of his birth, he was not
lacking in the virtues that became it. He was a good
and faithful friend: if a letter of commendation would

[1] *E*. vii. 18. 2.   [2] *E*. vii. 14. 10–12.   [3] *E*. ii. 9. 10; iv. 15.
[4] *E*. ii. 9. 2; cf. iv. 19.

assist an acquaintance in difficulties, he did not hesitate
to write;[1] the success or misfortune of his friends would
draw a letter from him even when his own circumstances did not encourage correspondence.[2] The per-
sistent support which he gave to Arvandus on his trial
procured him, as we shall see,[3] some unpopularity at
Rome—and that was no light thing for a man who
liked to see about him the smiles of friends.[4] 'To be
vanquished in affection', he says, 'is an abomination,'[5]
and though the phrase is not his own he seems to have
been true to it.

As has been explained, he does not give us much
information about his family life. He seems, however, to have been a good husband and father.
Between him and his family there reigned, he tells us,
a perfect concord: 'one might almost suspect,' he says,
'that we were charmed.'[6] Since their marriage his wife
bore him a son, Apollinaris,[7] and probably three
daughters, Severiana, Alcima, and Roscia.[8] He him-

---

[1] See *E*. ii. 4; ii. 5; iii. 9, &c.

[2] See i. 4; ii. 3; iii. 2; iii. 4; iv. 8; iv. 11, &c.

[3] *E*. i. 7. 1.   [4] Cf. *E*. i. 11. 17.

[5] *E*. ix. 11. 8 (= Pliny, *Ep*. iv. 1. 5).   [6] *E*. ii. 2. 3.

[7] Apollinaris was still studying at the time of the publication of the ninth
book (*E*. ix. 1. 5), but was old enough to make a journey to Rome very soon
afterwards. As will be shown later, this journey is to be dated to *c*. 479 or 481,
and thus Apollinaris will have been born *c*. 459–61 (see next note).

[8] The names of Sidonius' daughters present a difficulty. Severiana (*E*. ii.
12. 2) and Roscia (*E*. v. 16. 5) are both mentioned in the letters, but a third
name, Alcima, is given by Gregory of Tours (*Hist. Franc.*, iii. 2; iii. 12; *de Gloria
Mart.*, 64), who does not mention the other two. According to Mommsen (*ap.*
Lütjohann, p. 435; cf. p. xlix): 'Tria haec nomina num fuerint trium mulierum
uel duarum uel adeo unius, ignoratur'. We can, however, as I believe, go
farther than this. Two of the children were twins, and a poem written for the
occasion commemorates their sixteenth birthday (*C*. xvii. 3). As this poem
cannot have been published after 469, the year in which, as will be shown,
Sidonius became bishop of Clermont (see *E*. ix. 12. 1; ix. 16. 3, vv. 49–50, and
infra, pp. 205–7), it follows that if, as I have tried to show in the last note, Apollinaris was born *c*. 459–61, the twins of *C*. xvii. 3 cannot have included him. Unless

self superintended his son's education[1] and counselled him to avoid evil companions:[2] he was eager that the lad should have a distinguished career,[3] and was dismayed at his idleness.[4] When his daughter Severiana was stricken by a fever he showed himself a tender and affectionate parent.[5]

His moral standard seems to have been high.[6] Not a word of scandal has been breathed against him. With bad language he had no sympathy; evil speaking and virtuous living, he tells his son, are very rarely allied.[7] The frescoes in his baths at Avitacum were innocent, he tells us, of any indecency:[8] the seduction of a slave-girl on his estate was a 'wicked crime' sufficient to make him abandon his friendship with the seducer's master.[9]

On the other hand, his standards did not in general transcend those of his age. He had the aristocratic pride in his birth, which made him, when he had become a bishop, turn to St. Luke to justify it,[10] and with this he had the aristocratic disdain for the humbler classes. 'One can make allowances,' he says in one of his letters, 'for the uncultured;'[11] but it is just what this Gallic senator did not do. A friend of his who had become entangled with a slave-girl is congratulated by

there were children of Sidonius of whom we know nothing, we must assume that the twins were daughters. But neither of them can well have been Roscia, for in a letter which can be certainly dated to 474, we learn that she is still receiving instruction and that her age is tender (*E.* v. 16. 5, 'Roscia . . . cum seueritate nutritur, qua tamen tenerum non infirmatur aeuum sed informatur ingenium'). It is not at all probable that this language could be used of a girl born twenty-one years before. If this inference is correct, we may reject the hypotheses of Mommsen (l. c., p. xlix) and Germain (p. 6, n. 1) that there were one or two daughters, and say that the twins of *C.* xvii. 3 must have been Severiana and Alcima. Roscia and Apollinaris will be, on this view, younger.

[1] *E.* iv. 12. 1. [2] *E.* iii. 13. [3] *E.* v. 9. 4; v. 16. 4.
[4] *E.* ix. 1. 5. [5] *E.* ii. 12. 2. [6] Cf. *E.* iv. 9. 1.
[7] *E.* iii. 13. 11. [8] *E.* ii. 2. 6. [9] *E.* v. 19. 1.
[10] *E.* vii. 9. 17. [11] *E.* ix. 14. 8.

Sidonius on having abandoned her and taken a legal wife.[1] When the seducer was an aristocrat there is no longer a word of 'wicked crime'. Sidonius was a bishop when he wrote this letter, but there is no hint in it that the girl who had been betrayed and deserted had a claim upon the honour of her seducer. It is no great reproach to Sidonius that the aristocrat forgot what the bishop should have remembered. He lived in an age in which the saintly Augustine abandoned the woman who had been his first love,[2] in which Christian legislators declared that to punish a tavern-keeper's slave for adultery was futile because she was already so vile.[3]

This is not the only occasion on which Sidonius displayed this disdain and intolerance for those beneath him. He was always delighted to encourage the ambition for official distinction in his friends:[4] 'To an enlightened mind,' he says,[5] 'none seems nobler than he who steadily devotes himself and his resources to the single end of excelling his forefathers.' Yet how bitterly he condemns the attempts of the *novus homo*, Paeonius, to better his position,[6] and how violently he inveighs against the members of town councils who court popularity by doing official business without fee.[7] It was this aristocratic disdain which made him refuse to speak to a messenger who had lost the letters that he was carrying,[8] which made him assault the grave-diggers who had probably quite unwittingly defaced his grandfather's tomb.[9]

A man who was superior to the conventions of his

---

[1] *E.* ix. 6.   [2] See Stewart in *C.M.H.*, i, pp. 596–7.
[3] *Cod. Theod.*, ix. 7. 1.   [4] See *E.* i. 4. 1; iii. 6. 2.
[5] *E.* viii. 7. 3; cf. v. 18.   [6] *E.* i. 11. 6.   [7] *E.* v. 20. 2.   [8] *E.* iv. 12. 2.
[9] *E.* iii. 12. 3, '. . . more maiorum reos tantae temeritatis iure caesos uideri'.
On this passage see Allard, p. 75, n. 4; Dalton, ii, p. 227, and Coville, p. 54.

time would not have acted in this way, but Sidonius was not such a man. He followed out the conventions of his age to the end whether in action or in writing. Within his conventional framework he was apt to be an enthusiast; but he was not a deep thinker. He had small power of aesthetic criticism; and it has been pointed out that his observations on art are almost entirely borrowed from Claudian.[1] In philosophy, he remembered the tags of his school-days, and could sum up in a few lines of verse the main teaching of many thinkers,[2] but we scan his works in vain for any really comprehensive knowledge of philosophical thought. And in the conduct of his life his lack of depth of thought is equally shown. A cause such as the independence of Auvergne from the Visigoths or the security of the Catholic Church he defended with a courage and pertinacity to which we must give all our admiration, but we must not forget that he quite failed to read the signs of the times and to see that the future of Gaul lay in the establishment of a *modus vivendi* with the barbarians. As a bishop he tried to do his duty, as it lay before him, and he did it well; but he showed no power of deep speculation and no feeling for the equality of all men before God. As a writer he pursued the cult of antithesis and paronomasia farther than one would have believed it possible, he wrote much, and had a real enthusiasm for the company of learned men,[3] but there is barely a sentence in his collected works that shows evidence of hard independent thinking. What he found to his hand, he did indeed with all his might, but he did not look far.

[1] Purgold (*Claudianus und Sidonius*, 1878) quoted by Dalton (i, p. ci).
[2] C. ii. 165–81; xv. 44–195; cf. Mommsen, *Reden und Aufsätze*, pp. 137–8.
[3] E. vii. 14. 10.

V

SIDONIUS APOLLINARIS PREFECT OF THE CITY

WE have already noticed the complaints which Sidonius made to Gallic nobles on their reluctance to enter the public service, and when we remember that Sidonius himself was living in retirement for six years we may well ask whether such complaints came appropriately from his mouth.[1] Nevertheless, though there are general explanations for this reluctance which touch the whole history of the decline of the empire, the particular circumstances of these years 461–7 were such as to induce the aristocracy to remain on its estates rather than to pursue a career of public service.

The political situation during these years was to a Gaul more than usually confusing. The imperial throne was occupied by Libius Severus, but he was a mere puppet of Ricimer:[2] not an independent action of his is recorded by the chroniclers; that he was a Lucanian and that he reigned for six years living a pious life[3] are the only facts that we know about this phantom emperor. But the mere fact that there was a Roman emperor on the throne kept the Gauls for the most part loyal. There can be no stronger argument

---

[1] Mommsen (*ap.* Lütjohann, p. xlviii) and Mohr (p. 387) assume from *E.* iv. 14. 2 and 4 that Sidonius had been praetorian prefect of Gaul. The two passages, however, are inadequate to support the weight of an inference to which the complete silence of Sidonius in the rest of his works is opposed. Cf. Sundwall, pp. 133–4, and Stein, p. 546, n. 3.

[2] It is noteworthy that his coins bear on their reverse the monogram of Ricimer (Cantarelli, p. 78).

[3] *Chron. Gall.*, a. dxi., 636 (i, p. 664); Cassiodorus, *Chron.*, 1274 (ii, p. 157); *Laterculus Imperatorum*, iii, p. 423.

against a theory of Gallic separatism than the fact that throughout the reign of this creature of Ricimer's no attempt was made to set up a rival emperor in Gaul. Subsequent events, indeed, would lead us to believe that many of the Gauls were beginning to realize that their loyalty to the idea of a homogenous empire represented by a sovereign at Rome was out of touch with present realities, but throughout the reign of Severus such feelings were never expressed in action.

Yet there was one man who refused to acquiesce in the new régime: Aegidius, who had been rewarded for his services in 458 with the position of *magister militum per Gallias*,[1] had been a personal friend of Majorian, and the death of his master at the hands of Ricimer was an action that he could not forgive. He planned, as it appears, to march on Rome,[2] and it was only an attack made upon him by the Visigoths that forced him to abandon the attempt. Whether his project is to be understood as a refusal to recognize Severus as emperor, or whether it is anything more than the prosecution of a personal quarrel against Ricimer, the fragmentary nature of our authorities does not permit us to say. In either case Aegidius was actually a rebel and it is possible that the appointment of the Burgundian king Gundieuc as *magister militum* was made with the deliberate object of superseding him.[3] Since the Visigoths under Theodoric were still federates, their unsuccessful campaign against Aegidius in 463[4] is not a campaign

---

[1] Greg. Tur., *Hist. Franc.*, ii. 10 (11); cf. Tammasia, pp. 13–14.

[2] Priscus, fr. 30 (*F.H.G.*, iv, p. 104).

[3] So Stein, i, p. 504; cf. Thiel, *Ep. Pont. ad ann.* 463, p. 146 (= Hilary, *ep.* 10), and Ensslin, *Klio*, xxiv (1931), pp. 491–2.

[4] For which see Hydatius, 218 (ii, p. 33); *Chron. Gall., a. dxi.*, 638 (i, p. 664); Marius Avent., ii, p. 232; Greg. Tur., *Hist. Franc.*, ii. 13 (18).

against the empire at all. Aegidius is not a Roman general fighting against barbarians hostile to the empire,[1] but a rebellious subject attacked by loyal federate troops, and this is shown clearly from the fact that in 464 he is found negotiating with Geiseric,[2] the empire's most implacable foe. Aegidius, in fact, and his son and successor, Syagrius, are rulers in complete independence of the empire.[3]

The curious state of affairs, which made the campaigns of a barbarian king against a former *magister militum* of Gaul an act of loyalty to the empire, are strangely illustrated by the incident of Narbo in 462. It is related[4] that the Count Agrippinus, who was a personal enemy of Aegidius, in order to gain the assistance of the Visigoths handed over to them the town of Narbo. The assistance that he demanded can only have been needed against Aegidius, and we may suspect that the transfer of Narbo was the price exacted by Theodoric for the campaign of 463. The inhabitants of Narbo were not unnaturally bewildered at the complicated diplomatic design for which they had been sacrificed, and resisted the Visigothic occupation to the best of their ability; but their efforts were unavailing.[5]

Such a strange situation would seem barely credible if it were not supported by the clear evidence of contemporaries. We have a letter written by Hilarus,

---

[1] It is significant that there is no mention of Aegidius in Sidonius' works, for the theory that the unnamed general of *C*. v. 553–7 is to be identified with him has been refuted by Cessi, *Ati di Reale Ist. arch. Veneto* (1917), p. 1120. Sundwall (p. 14) alleges that Aegidius was supported by 'Der gallische Hochadel'. But we have no evidence for this and (as will be seen) we have evidence against it. [2] Hydatius, 224 (ii, p. 33).

[3] Ib. 228 (ii, p. 33); cf. Tammasia, pp. 36–7.

[4] Hydatius, 217 (ii, p. 33), Cantarelli, p. 81. The account of *Vita S. Lupicini*, 11–14 (*M.G.H., Scrip. Rer. Merov.*, iii, pp. 149–52) is justly rejected *in toto* by Schmidt (i, p. 258, n. 1). [5] *C*. xxiii. 59, 74–87; cf. *C*. xxii, ep. 1.

bishop of Rome, to Leontius, bishop of Arles:[1] this letter is dated to November 3rd, 462, almost certainly after the seizure of Narbo. So far from expressing any note of regret or consolation at the event, Hilarus actually praises the 'magnificence' of Frederic, brother of the Gothic king, and calls him 'his son'. Again, very soon after the capture of Narbo, Sidonius in his poem to Consentius, who was himself an inhabitant of *Narbonensis prima*, describes Theodoric in the warmest manner as a faithful defender of the empire.[2] Nevertheless, when they realized that loyalty to the empire involved acquiescence in the advance of barbarian kings to east and to west, when they saw Theodoric possessed of Narbo and Gundieuc ruling at Lyons,[3] we can well believe that for many of the Gallic nobles the feelings of loyalty which bound them to the empire became weaker. 'The state,' said Sidonius, 'seems to care nothing for its most devoted sons, but can you wonder, when a race of uncivilized federates is directing or rather destroying the Roman power?'[4]

---

[1] Thiel, *Ep. Pont. Rom.*, i, p. 140 (= Hilary, *Ep.* 8).

[2] *C.* xxiii. 70-2. 'Magno patre prior, decus Getarum, / Romanae columen, salusque gentis, / Theudoricus . . .' Sundwall's statement (p. 14) that the Goths were 'die Feinde des römischen Galliens' is directly contradicted by this passage (which he does not quote); and his whole theory of an opposition between Italy (under Ricimer) and Gaul (under Aegidius) is seen to be unjustifiable.

[3] The date at which the Burgundians occupied Lyons cannot, it is true, be certainly fixed. It must be at some time between 461 and 474 (*E.* vi. 12. 3). It seems plausible to suppose that the friendship of the Burgundians with Ricimer was purchased like that of the Visigoths by some definite concession. If the date (461) ascribed to *C.* xii is correct (see above, p. 66), it gives a slight support to the view that the Burgundians were in possession of Lyons quite soon after 461, and it certainly appears from Thiel, *Ep. Pont. Rom.*, i, p. 148 (= Hilary, *Ep.* 10), that Vienne and Die were in Burgundian territory by the end of 463 (cf. Schmidt, i, p. 375).

[4] *E.* iii. 8. 2. There is, it is true, no definite evidence for assigning a date to this letter, but its tone seems to suit the circumstances of this time.

The death of Severus at Rome in November, 465,[1] made little difference to the political situation: the nominal ruler of the whole empire was again Leo, but as before the real direction of affairs in Italy was in the hands of Ricimer. It is noteworthy that now there was nothing like a 'coniuratio Marcelliana' among the nobles of Gaul: they had begun at last to despair of the republic.

Theodoric, as will be remembered, had ascended the throne by the murder of his brother, and now after a reign of fifteen years he fell by the same fate. In 466 he was assassinated at Toulouse by his brother Euric, who succeeded him.[2] The new king was one of the most conspicuous figures of the century; even his opponents confessed the force of his personality and strength of will.[3] A fanatical Arian, who believed, like Geiseric, that he owed his success to the direct aid of God;[4] he did not, however, let his religion interfere with his statesmanship. Moreover, if his servants did their work well, he cared not whether they were orthodox or Arian. He enlisted the best intelligence of his Gallo-Roman subjects into his service,[5] and they repaid him with fidelity. The pre-eminent characteristic of his statesmanship was its political realism. Euric very soon saw that the position of an imperial federate, especially when it involved dependence upon a barbarian in Italy, had nothing to be said in its favour. He did not

[1] For the date, see Seeck, *Untergang*, vi, pp. 483–4, correcting his statement in ib. 352.

[2] Hydatius, 237, 238 (ii, p. 34); *Chron. Gall., a. dxi.*, 643 (i, p. 664); *Chron. Caesaraug.*, ii, p. 222; Marius of Avent., ii, p. 233; Jord., *Get.*, xlv. 235. For the date see Yver, p. 14.

[3] See *E.* vii. 6. 6; viii. 6. 16; viii. 9. 5; Ennodius, *Vita Epiphanii*, 86.

[4] *E.* vii. 6. 6.

[5] As Leo, Namatius, and Victorius (cf. Seeck in *P.-W.*, vi, p. 1239).

choose the course which Theodoric had followed; he did not, as it appears, wish to set up an emperor to rule under his protection. His model was Geiseric the Vandal; he wished his kingdom in Gaul to be completely independent of the empire. Even such an indifferent historian as Jordanes saw clearly that his reign inaugurated a change of policy. 'Euric's object,' he said, 'was to rule Gaul in his own right.'[1] Such a policy, therefore, compelled every Gallo-Roman inhabitant to declare himself definitely for Euric or the empire. A feeling of vague loyalty towards the imperial idea was now no longer possible. Euric had thrown down a challenge, and the history of the next ten years in Gaul is nothing but a record of the way in which that challenge was met.

On his accession Euric at once dispatched an embassy to Leo at Constantinople,[2] and it has been conjectured with much plausibility that he was already seeking to obtain a modification of the *foedus*.[3] A convention was made with the Suevi, and an embassy was also sent to Geiseric.[4] The conversations of these ambassadors are not recorded to us: nevertheless, we cannot but suspect that an embassy to Geiseric was a blow directed at the empire.

Meanwhile in Italy there had been important political changes. The continual depredations of the

---

[1] *Get.*, xlv. 237; cf. *E.* vii. 6. 4. F. Lot (*La Fin du monde antique et le début du Moyen Âge*, p. 286) rejects the evidence of Jordanes and Sidonius on the ground that Visigothic coins continue to carry the head of the emperor (Hess, *Monaies des Rois wisigoths* (1872), pp. 28–9). But we can well believe that a government not too sure of itself would be unwilling to tamper with the monetary unit: moreover, Wroth (*Coins of the Vandals, Ostrogoths, &c. in the British Museum*, p. xxix) shows that the Ostrogothic kings kept the heads of Byzantine emperors on their coins, even when they were at war with the empire.

[2] Hydatius, 238 (ii, p. 34).   [3] Schmidt, i, p. 260.

[4] Hydatius, l. c.; Yver, p. 18.

Vandal pirates along the coasts of Italy were too much for Ricimer's unaided efforts:[1] he needed the support of the Eastern Empire. But Leo was by no means inclined to help except at his own price;[2] he demanded that the next emperor of the West should be chosen by the East, and Ricimer was forced to comply. Leo's choice fell upon Anthemius, the son-in-law of Marcian,[3] his predecessor on the throne of Constantinople. He was sent to Italy with a large body of troops and declared Augustus outside Rome (April 12th, 467).[4]

The main object of the new emperor was to crush the naval power of the Vandals,[5] but he desired also to make more close the contacts between the Roman state and the Gallic aristocracy, and, if possible, to check the designs of Euric. An alliance was made with the independent Bretons, north of the Loire, as a result of which their king Riothamus undertook with a force of 12,000 men to defend Berry against the Visigoths.[6] Friendly relations were restored with Syagrius in the north-east, and as a result of this the Franks appear again as federate barbarians fighting for the empire:[7] the Burgundians had, indeed, remained faithful to the empire

---

[1] *C.* ii. 353–5, 381–2.

[2] Seeck (in *P.-W.*, ii, 2, p. 2006; *Untergang*, vi, pp. 352, 483), following Jordanes, *Rom.*, 335, denies that Severus had ever been recognized by Leo, but he is refuted by Cantarelli, p. 78, n. 3; cf. Baynes in *J.R.S.* xvi (1922), pp. 222–3.

[3] For the family tree of Anthemius see Hodgkin (ii, p. 451).

[4] *Fast. Vind. Priores*, 598 (i, p. 305) (date); Hydatius, 234, 235 (ii, p. 34); Marcellinus Comes, ii, p. 89; Cassiodorus, *Chron.*, 1283 (ii, p. 158); Theophanes, *ad. a.m. 5958*; Peter Patricius, *ap.* Constant. Porph., *de Cerem.*, i. 87.

[5] *C.* ii. 14, 18.

[6] Jord., *Get.*, xliv. 237; cf. *E.* i. 7. 5.

[7] Count Paulus and the Franks are represented as fighting for the Romans against the Visigoths near Angers *c.* 469 (Greg. Tur., *Hist. Franc.*, ii. 13 (18)). The marriage of the Frankish prince, Sigismer, to a Burgundian princess may date from this time (*E.* iv. 20; cf. Coville, p. 72).

for the last ten years, and their king, Gundobad, who seems to have succeeded his father Gundieuc about this time, was, like him, granted the position of *magister militum*.[1] Measures of conciliation were also applied towards the Gallic aristocracy, and Sidonius, who had been sent by the Arvernians to lay a petition[2] before the emperor, was assisted in every manner on the journey and was granted the privileges of the imperial post.[3] He was in an optimistic mood as he set out the second time for Rome, and was delighted at the opportunity of revisiting the city. It almost seems as if the mere fact of having seen Rome was enough to instil into the mind of a provincial a passionate devotion towards the imperial idea. One would hardly expect a man to have the kindest memories of a city whose inhabitants had ejected his father-in-law with contumely, but the emotions which Sidonius felt at the name of Rome transcended any bitterness that he might feel against her people. To a Gallic friend he writes in tones of remarkable warmth:

'I appeal to you,' he says, 'to enter the civil service. . . . For then a man may visit Rome once in his prime; Rome, the abode of law, the training school of letters, the fount of honours, the centre of the world, the mother-land of freedom, the city unique upon earth, of whom all are citizens save the slave and the barbarian.'[4]

Sidonius has left us a spirited account of his journey.[5] His route went through Lyons, over the Alps, down

---

[1] John Malalas, xiv (Migne, *Patr. Graec.*, xcvii, p. 557); Schmidt, i, p. 375. On the relations between the empire and the Burgundians see Jord., *Get.*, xlv. 238 and *E.* v. 5; v. 6; cf. Sundwall, p. 15.

[2] *E.* i. 9. 5, 'Aruerna petitio'. We know nothing of its content.

[3] *E.* i. 5. 2. [4] *E.* i. 6. 1–2.

[5] *E.* i. 5.

the Ticino and Po to Ravenna. From there he went south to Ariminum and so to Rome by the Flaminian way. During the journey he suffered from a sharp attack of fever, and perhaps for that reason in his description of the journey he hurries over the latter part across the Apennines. One can see clearly how the interest of a man, whose earliest studies had been mythology and ancient history, was aroused when he passed by the sites where the sisters of Phaethon had been transformed or the battles of the Punic War had been fought.

Some time towards the autumn of 467[1] Sidonius approached for the second time the city of Rome. With the fever still heavy upon him, he did not at once enter the walls, but made a detour around them to the churches of the apostles, Peter and Paul, to whom he prayed for relief. His prayer was successful, for the fever suddenly left him:[2] he entered the city and took up his quarters at an inn.

The occasion was not very propitious for a suppliant at the Imperial Court. Sidonius' arrival, as he tells us, coincided with the nuptials of Ricimer and Alypia, the daughter of Anthemius:[3] all business was suspended and the city abandoned itself to merry-making.

Sidonius made use of the time at his disposal to ensure the success of his mission by intrigue. He had now made the acquaintance of Paulus, a cultured Roman who had been twice prefect of the city, and had vacated

[1] The petition was still unheard when Sidonius was urged to write the panegyric to Anthemius shortly before Jan. 1st, 468 (E. i. 9. 5), so that he could not have arrived in Rome much earlier.

[2] E. i. 5. 9, 'post quae caelestis experimenta patrocinii . . .'. If it were not for one's knowledge of the habitually ponderous language of Sidonius, such a phrase would almost seem to suggest scepticism.

[3] Ib. § 10, 11; cf. i. 9. 1; John Ant., fr. 209 (F.H.G., iv, p. 617).

his lodgings at the inn to enjoy the hospitality of his new friend. With him he discussed the best means to gain the favour of the court.[1] It appeared that there were only two senators whose influence would be valuable, Gennadius Avienus and Caecina Basilius. They were both men with distinguished records in the public service. Avienus, the elder of the two, had been consul in 450 and had accompanied Pope Leo on the embassy to Attila in 453. Basilius had been twice consul and twice praetorian prefect of Italy.[2] It is interesting to note that the patronage of these two senators was among other reasons the more desirable because it could be obtained without any great expenditure of money.[3] As between the two of them, Avienus was the more specious, but Basilius the more solidly valuable protector. The whole incident as recorded by Sidonius, with its assumption that intrigue was essential for the success of his mission,[4] throws a curious light upon the relations between the provinces and the central government at this period.

Basilius showed himself a useful friend; he suggested to Sidonius that he should compose a panegyric in honour of the approaching consulship of Anthemius, and promised to use his influence with the emperor to secure its delivery.[5] In this he was successful, nor can we doubt that Anthemius was quite prepared to entrust the panegyric to a Gallic orator of approved reputation. Such a concession would be to display in a practical fashion his sympathy with the Gallo-Roman aristocracy.

[1] E. i. 9. 1–4.   [2] Sundwall, pp. 54, 55, 56.
[3] 'neuter aditu sumptuoso', E. i. 9. 4.
[4] 'igitur per hunc primum, si quis quoquo modo in aulam gratiae aditus, exploro', E. i. 9. 1.   [5] E. i. 9. 5–6.

On January 1st, 468, Sidonius delivered the last of his public panegyrics. Like its immediate predecessor it was written in extreme haste,[1] and in writing to a friend at Lyons, to whom he sent a copy of it, he pleads for an indulgent judgement.[2]

The panegyric is a hexameter poem of 548 verses, preceded, as usual, by a short elegiac introduction. It resembles in general structure the earlier panegyrics, but as the author knew far less about the career of Anthemius than about his previous heroes, there is more padding than ever before. A long dissertation on the various authors that Anthemius might be supposed to have read occupies nearly forty lines,[3] a careful inventory of the cave of Aurora, followed by a speech put into the mouth of Rome, in which she recapitulates the history of her relations with the East since the time of Mummius, occupies another seventy.[4] Here alone we find one-fifth of the poem devoted to matter which has only the slightest relevance to the main theme. And, owing to haste, though the material of the various disquisitions is skilfully managed, they are not fitted into a coherent whole. The poem is, in fact, too obviously a piece of patchwork. Even Sidonius' facility in constructing crisp, epigrammatic phrases

---

[1] In describing the circumstances of the poem's composition, Sidonius' language (*E.* i. 9. 5–6) is obscure, 'ilicet, dum de . . . petitionibus elaboramus, ecce et Kalendae Ianuariae, quae Augusti consulis mox futuri repetendum fastis nomen opperiebantur. Tunc patronus: "heia", inquit, ". . . exeras uolo in obsequium noui consulis ueterem Musam uotiuum quippiam uel tumultuariis fidibus carminantem" '. From this it looks as though the conversation with Basilius took place on Jan. 1st and that Anthemius assumed the consulship at a later date ('mox futuri'). On the other hand *C.* ii. 3 shows clearly that the panegyric was delivered on Jan. 1st. Perhaps Sidonius has made a slip in his letter.

[2] *E.* i. 9. 7.   [3] *C.* ii. 156–92.

[4] Ib. 407–77.

seems to have deserted him.[1] Of his three public poems we can certainly say that this is the dullest.

The main theme is the appeal of the West to the East, and the dispatch of Anthemius in answer to it. With the two emperors reigning in such concord, declared the poet, the empire would be one, and their united strength would deliver the Mediterranean lands from the Vandal piracy that had so long beset them.[2] That was a wish in which every imperial subject, whether in Gaul or Italy, might concur.

Now that the marriage between Ricimer and the daughter of the emperor had so recently taken place, Sidonius could hardly omit his name from the poem, or content himself with the significantly slight reference that he had given in the panegyric to Majorian. To praise the man who had been directly responsible for the destruction of Avitus and Majorian must have been difficult, even for Sidonius, but to escape it was barely possible. It cannot, at least, be said that his praises are not strictly accurate and indeed deserved: the royal birth of Ricimer, and his victories over the Vandals and Alani[3] are indisputable facts, and if we cannot prove that his might struck fear into the hearts of barbarians on the Rhine and in Noricum,[4] it is quite probable in itself. On the other hand, listeners might have detected in several passages of the panegyric an implied disapproval of Ricimer and the policy for which he stood. The deliberate manner in which the poet introduces the statement that the emperor Severus had died a natural death[5] must surely proceed from a man who

---

[1] 'conuersoque ordine fati / torrida Caucaseos infert mihi Byrsa furores' (350, 351), is one of the few notable sentences in the poem.

[2] Ib. 16–17, 28–9.   [3] Ib. 360–70.   [4] Ib. 377–80.

[5] Ib. 317–18, 'auxerat Augustus *naturae lege* Seuerus / diuorum numerum'.

was acquainted with the rumour that he had been murdered by Ricimer.[1] And it was barely tactful to say in his presence that the republic would be better governed now that there was a worthy emperor.[2] Still more pointed was the remark that the country's real need was an emperor who would not authorize but lead expeditions.[3] The poet may have been right; but such an emperor was certainly not desired by Ricimer. In fact his main object had been to ensure that the emperor should be too weak to do anything more than authorize.[4] The words, indeed, were an indictment of Ricimer's policy; one can hardly believe that they failed to find their mark.

Whatever we may think of the panegyric, the poet was brilliantly rewarded for it. He was nominated by the emperor to the office of *Praefectus Urbi*. Sidonius at once wrote a letter to his friend Herenius at Lyons, announcing the good news. 'With Christ's aid,' he wrote, 'I have gained the prefecture by my lucky pen.'[5] He enclosed a copy of the work which had brought him fortune, and shortly afterwards, as it appears, he published the complete collection of his official poems. In compliment to the emperor, the chronological order is inverted so that the panegyric to Anthemius comes first in the collection.[6] Sidonius might indeed congratulate himself upon the worldly success that his poetry had

---

[1] Cassiodorus, *Chron.*, 1280 (ii, p. 158).

[2] *C.* ii. 15–16, 'respublica . . . digno melius flectenda magistro'.

[3] Ib. 382–4, 'modo principe nobis / est opus armato, ueterum qui more parentum / non mandet sed bella gerat'.

[4] Cf. Bury, i, p. 341.   [5] *E.* i. 9. 8.

[6] This is the very plausible conjecture of Germain (p. 40). The other poems, as we shall see (p. 108), were published at a later date from Auvergne, and Leo (*ap.* Lütjohann, p. xli) gives reasons for supposing that the public poems originally formed a separate collection.

won for him: we may, however, believe that his pro-
motion was due at least as much to the emperor's wish
to conciliate the Gallic aristocracy as to the merits of
the panegyric.[1]

Though the Prefect of the City was no longer next
in rank to the praetorian prefects below the emperor,[2]
since the higher military officials now took precedence
of him,[3] the office was still of great importance.[4] The
praefect of the city was a *vir illustrissimus*, and enjoyed
the right of travelling in a four-horse chariot.[5] He was
the head of the administration of justice and of police
not only in the city of Rome,[6] but over a radius of a
hundred miles around it;[7] and thus portions of the provinces of Tuscia, Umbria, Campania, and Picenum[8]
were under his control. From his tribunal there was
a right of appeal to the emperor alone.[9]

If not more important, at least more irksome, was
his position as the official ultimately responsible for
the maintenance of the supply of corn, wine, and other
commodities, which were provided either free or at a
very reduced price for the consumption of the people.
The actual details of this supply service were regulated
by his subordinates, the *Praefectus Annonae*,[10] the Count
of the Port, the Registrar of Wine, and the Tribune of
the Pig-market,[11] but many passages in the literature of

---

[1] So Baret, pp. 23–4.
[2] As in *Not. Dig. Occ.*, i (ed. Seeck, p. 103); cf. *Cod. Theod.*, vi. 7. 1.
[3] See *E.* i. 9. 2, and Stein, p. 563, n. 1.
[4] On the office of *praefectus urbi* see Reid in *C.M.H.*, i, pp. 50–1; Stein, pp. 63–4.
[5] *Not. Dig. Occ.*, iv (ed. Seeck, p. 113), Cassiod., *Var.*, vi. 4. 6.
[6] *Cod. Theod.*, i. 6. 7; ix. i. 9; Symmachus, *Ep.* x. 36; S.H.A., *Vita Marci*, xi. 9.
[7] *Dig.*, i. 12. 1. 4; *Cod. Theod.*, ii. 16. 2; Cassiod., *Var.*, vi. 4. 5.
[8] See Godefroy ad *Cod. Theod.*, xiv. 6. 1.
[9] *Cod. Theod.*, xi. 34. 2.   [10] *Cod. Theod.*, i. 6. 5.
[11] *Not. Dig. Occ.*, v (ed. Seeck, p. 114), 'Comes Portus, Rationalis Vinorum,
Tribunus Fori Suarii'.

the period show us how a scarcity of food was visited on the head of the prefect himself.[1] At the public spectacles he had the difficult task of keeping order and preventing riots.[2]

Not only was the prefect of the city the chief magistrate of Rome, he was also president of the Roman senate.[3] He summoned the senators to meetings,[4] and was the first to address them.[5] He was, as a general rule, the judge of civil actions in Italy in which a senator was defendant,[6] and in the event of a criminal charge against a senator he sat as president of a commission composed of five of his colleagues, chosen by lot, to try the case.[7]

Seeing that Sidonius was praefect of the city in 468, a year conspicuous for the great naval attempt of the two empires against Geiseric, it is a matter of disappointment that he gives us practically no information about his tenure of the office: it almost seems as if he felt that this year of his life was not for him a glorious success. He held the office indeed in difficult times. With Africa in the hands of an unfriendly power, the danger of a failure in the supply of provisions must have been very great.[8] Indeed, the only letter which we can certainly assign to this year bears eloquent testimony to his difficulties. In writing to a friend who had given the

---

[1] Cf. Amm. Marc., xv. 7, 3; Symm., *Ep.* ii. 6; vi. 18.

[2] Cassiod., *Var.*, i. 32. 1; vi. 4. 6.

[3] *E.* i. 9. 6; Cassiod., *Var.*, i. 42. 3. It appears from *C.I.L.*, vi. 1698, that this was not always so. Cf. Lécrivain, p. 59, n. 10.

[4] Cassiod., *Var.*, ix. 7. 6.   [5] Ib. vi. 4. 3.

[6] *Cod. Theod.*, ii. 1. 4; *Dig.*, i. 9. 11; *Cod. Iust.*, iii. 24. 2. According to the rule: 'actor rei forum sequitur', if a senator is plaintiff the case is judged by the local governor (cf. *Cod. Iust.*, iii. 22. 3; Lécrivain, pp. 91–2).

[7] *Cod. Theod.*, ii. 1. 12; ix. 1. 13; Cassiod., *Var.*, iv. 22. 3.

[8] On the importance of the African corn-supply see Salvian, *De Gub. Dei*, vi. xii. 68, and vii. xiv. 60.

praefectus annonae an introduction to him, he writes that there may be an uproar in the theatre if the corn does not arrive, and that he will be held responsible for the resulting famine. 'Fortunately,' he writes, 'five ships from Brundisium have put in at Ostia with cargoes of wheat and honey, and I am sending the *praefectus annonae* to arrange for their distribution.'[1]

It was perhaps fortunate that Sidonius' term of office did not last longer than a year;[2] for he was thus able to avoid the unpleasant duty of having to preside over the commission that judged his friend Arvandus, the praetorian prefect of the Gauls, on a charge of treason (469). Arvandus[3] had been prefect of the Gauls for five years (464–8)[4] during which time his term of office had been once renewed.[5] His first term was successful and popular, but during the second he fell into debt and oppressed the provincials. For this the Council of the Seven Provinces ordered his arrest[6] and he was sent to Rome to be tried, as was his right, by a court of five senators over which the city prefect presided.[7] His accusers were Tonantius Ferreolus, Thaumastus, and Petronius, all three distinguished

---

[1] *E.* i. 10. 2.

[2] Dalton seems to consider (i, p. xxxiii) that Sidonius' withdrawal from office was a voluntary retirement, and advances reasons for it. They are not necessary, for examination of Sundwall's list of fifth-century city prefects (pp. 24–6) shows that a term of office lasting more than one year is exceptional.

[3] On the trial of Arvandus see Duval-Arnould, pp. 36–54; Carette, pp. 333–54. His identity with the Arabundus of Cassiodorus, *Chron.*, 1287 (ii, p. 158) (date), and the Servandus of *Hist. Misc.*, xv. 2, cannot be doubted.

[4] *E.* i. 7. 3, 11.

[5] *E.* i. 7. 11.

[6] Ib. § 4.

[7] *E.* i. 7. 9. The MSS. give 'xuiris', but since we have evidence for the trial of senators by a court of five immediately before and shortly after this date (see p. 102, n. 7), 'Vuiris', an emendation already known to Sirmond (pp. 16–17), should probably be substituted.

Gallic nobles.[1] It was an embarrassing position for
Sidonius; the prosecutors were all personal friends,
Ferreolus and probably Thaumastus were related to
him, but he had also close personal ties with the accused man. Arvandus' actions had prejudiced public
opinion against him in Rome, and his supporters came
in for a share of that unpopularity, nevertheless Sidonius felt that he must do his duty by him. He soon
learnt that his friend's case was worse than he had suspected; there was to be formulated an accusation not
only of extortion but of treason. A letter had been
intercepted from Arvandus' secretary, which he confessed under torture to have been written by his master.
It was addressed to the Gothic king, and urged him in
decisive language to make war on the 'Greek emperor'[2]
It further advised an attack on the Bretons who had
come to defend Berry and urged that Gaul should be
divided up according to the law of nations[3] between
the Visigoths and the Burgundians. Sidonius, who saw
how great would be the danger if such a treasonable
correspondence were brought home, begged Arvandus
to make no admissions whatever.

The accused, however, seemed quite confident that
he would escape. He trusted, he said,[4] in his own conscience, and would only allow advocates to defend him
on the charge of peculation. Meanwhile he walked
about Rome in complete assurance. His demeanour at
the trial was the same. He endeavoured to sit down on
the judges' bench, while his accusers, dressed in black,

---

[1] All three are recipients of letters from him. For their relationship see *E.*
v. 6. 1; v. 7. 1; and vii. 12. 1.

[2] *E.* i. 7. 5, 'pacem cum Graeco imperatore dissuadens'. Cf. Ennodius, *Vit.
Epiphanii*, 54 ('Graeculus').

[3] *E.* i. 7. 5, 'iure gentium'.   [4] Ib. § 6, 7.

sat quietly and modestly in the lowest seats. Not only did Arvandus make no attempt to deny the authorship of the incriminating letter, but he admitted it without waiting to be examined. The judges exclaimed that such an admission was equivalent to a confession of treason. He was convicted and taken to the common jail. A fortnight later sentence of death was pronounced upon him. A law of Theodosius I had allowed a lapse of thirty days between sentence and execution,[1] and during this time the friends of Arvandus did what they could to save him by petitions and prayers. Their efforts were successful, for his life was spared, but he was sent into exile.[2] He is never heard of again.

The trial of Arvandus is an event of great interest in the political history of the period, but it would be of still greater if we could clear up the mystery that surrounds it. The fact that a high Gallic official is prepared to support the extreme claims of Euric, and to consider that the law of nations demands a division of the country between two barbarian rulers is a phenomenon quite new in the history of Gaul. Even if it were true that the 'coniuratio Marcelliana' was a symptom of Gallic separatism, this is a proposal far in advance of it —a proposal which accepts the fact that the régime of a homogeneous empire must give way to that of separate barbarian kingdoms. The system of barbarian 'federation' had proved its own futility and unreality by the events of Narbo and the Loire in 462–3. Euric was realist enough to see this, and now Arvandus had seen it

---

[1] *E. i.* 7. 12, and Duval-Arnould, pp. 52–3.

[2] Cassiod., *Chron.*, 1287 (ii, p. 158); *Hist. Misc.*, xv. 2. Baret says (p. 26) that his exile 'probablement ne fut pas de longue durée'. How does he know?

too. Whether there is any connexion between the peculation and the treason of Arvandus we can hardly say,
but it is significant that no charges were made against
him until the second term of his prefecture, that is,
until Euric was on the throne. Significant too is the
fact that his projects were rejected by the majority of
the Gallic aristocracy:[1] loyalty to the imperial idea was
still strong.

Historians have not failed to notice and comment
upon the strange fact that Arvandus, though obviously
guilty of treason, expected up to the last moment that
he would be acquitted. Sidonius, it is true, has an
explanation. 'Arvandus,' he says, 'discovered only too
late that a man could be convicted for treason even if
he had not assumed the purple.'[2] It is not stated that
Arvandus actually used these words; Sidonius was told
by a friend that this was the reason for his perturbation.
One cannot but feel that it is unconvincing. Arvandus'
case is quite clearly covered by the opinion of jurists[3]
and by direct imperial legislation,[4] and one can hardly
believe that an official entrusted with the administration of the law should have been so ignorant of the
fundamentals of it.

More plausible is the theory[5] that the confidence
which Arvandus showed in the face of such damaging
charges was influenced by the fact that he had power-

[1] The phrase of Sidonius (*E.* i. 7. 3)—'successuros sibi optimates aemulabatur'
—is probably significant, but it is too narrow a base on which to construct a
hypothesis. The obscure words of ib. § 11—'plebeiae familiae non ut additus
sed ut redditus'—imply that Arvandus, like Paeonius, was a *novus homo.* Yver
(pp. 21–5) seems to believe that the pro-Visigoth party in Gaul represented a
movement of the lower classes against the aristocracy. We shall meet this view
again in the discussion of the siege of Clermont; it may contain some truth,
but the evidence for it is inadequate.

[2] Ib. 11.   [3] *Dig.*, xlviii. 4. 4.   [4] *Cod. Theod.*, ix. 14. 3.
[5] Tillemont, *Hist. des Emp.*, vi, pp. 347–8; Cantarelli, pp. 95–6.

ful friends in Rome from whom he expected an acquittal. Such friends could only exist among the party of Ricimer. If this supposition is correct, we must assume that Ricimer was a supporter of the plan which Arvandus had proposed. This is not at all improbable in itself. If Ricimer had reason to foresee a rupture with the emperor, it was in his interest that Gaul should be under barbarian control, and thus remain at least neutral in the struggle. That such a situation would have been to Ricimer's advantage is shown by the fact that in 472 Anthemius did actually receive assistance from Gaul.[1] Sidonius was certainly convinced that his client was totally lacking in judgement,[2] but unless we postulate the existence of some such intrigue with Ricimer, we must assume that he lacked sanity as well.

Sidonius had already left Rome before the trial of Arvandus had begun,[3] he had received the title of patrician[4] and could return to Gaul with the feeling that he had gained an honour greater than those of his forefathers. Only the consulship remained, but that honour fate was to place for ever beyond his reach.

---

[1] *Hist. Misc.*, xv. 4.

[2] 'Morum facilitas uarietasque. . . . non habuit diligentiam perseuerandi', *E.* i. 7. 1, 2.   [3] *E.* i. 7. 9.   [4] *E.* v. 16. 4.

VI

BISHOPRIC

SOON after his return to Auvergne early in 469, it appears that Sidonius published a book containing fifteen of his non-official poems;[1] and this, with the panegyrics already published, completes the tale of Sidonius' poems that are preserved in our manuscripts.

We must not, however, assume that in this series we possess all the poems that Sidonius ever wrote. Apart from the volume of his juvenilia, which he had published long ago and which is now lost, there are several short poems—epitaphs and occasional verses—preserved in the body of the Letters.[2] On one occasion, when he sends a poem in a letter to a friend, he says that it shall be added to his published collection, if the recipient thinks it good enough.[3] This poem, at least, was never included in the collection, and, like another, was laid up in Sidonius' bureau, 'to be gnawed by the mice'.[4] No doubt there are others, like the long poem to Eriphius

---

[1] These poems were certainly published before Sidonius became bishop (*E.* ix. 12. 1; ix. 16. 3, vv. 55–6; cf. *E.* iv. 3. 9); and, as has been shown, *C.* xvii. 3 gives a *terminus a quo* of 469 for the complete collection. Klotz, however (in *P.-W.*, ii, A. 2, p. 2233), points out that the dedication poem to Felix (*C.* ix) probably accompanied originally a shorter collection, for the words 'nos ualde sterilis modos Camenae / rarae credimus hos breuique chartae' are not very appropriate to the complete edition of sixteen poems. And the fact that the epitaph in *E.* ii. 8. 3 seems to have been written after the publication of at least one volume of 'epigrammata' lends support to this view. (On the meaning of 'Epigramma' in Sidonius see *E.* ix. 12. 3; *C.* xxii, ep. 6.) But though we may assume that this preliminary edition appeared *c.* 464–7, we cannot say definitely what poems it contained.

[2] See *E.* ii. 8. 3; ii. 10. 4; iv. 18. 5; viii. 11. 3; ix. 13. 5.

[3] *E.* ii. 8. 3 quoted above.

[4] *E.* ix. 13. 6.

on 'the man who bore good fortune ill',[1] which were
never published at all.

Sidonius showed an extreme modesty about the
merits of his poems. Their success, he says, is due more
to luck than to skill;[2] they are 'mere trifles',[3] 'only good
enough to be made into parcels for pepper or sprats'.[4]
Such modesty as this defeats its own end; a man who
writes in this strain of extreme self-depreciation is
surely a man who has an inward confidence in the
value of his own work. If we take him at his word,
however, we find that he has, perhaps unwittingly,
provided us with a true criticism. 'I write', he says,
'more with a ready than with a gifted pen';[5] and this
phrase is no less true for Sidonius than for the man from
whom he borrowed it. His poetry is above all things
fluent; it is never rough. No matter what metre he
essayed, he could always produce tolerable verse. His
poems, as far as one can judge, must have sounded well;
he has a good control over the hexameter, and in his
long panegyrics the position of the caesura is ingeniously
varied so that they escape monotony. His mastery
over the technique of metre was such that he could
and did write very fast. On the three impromptu pieces
of verse preserved in his letters,[6] we should be as kindly
critics as the unusual circumstances of them demand,
but the poem for the cup of Ragnahild[7] and the epitaph of his grandfather Apollinaris[8] were both written,

[1] *E.* v. 17. 11. Fertig (iii, p. 19) quotes *E.* ix. 16. 3, v. 39, 'Sapphico creber
cecini', and points out that no such Sapphics are now extant except the poem
from which this is quoted.   [2] *E.* i. 1. 4.

[3] *E.* ix. 13. 6; *C.* ix. 9; cf. *E.* ii. 10. 3-4.   [4] *C.* ix. 320.

[5] *E.* iii. 7. 1 (= Pliny, *Ep.* vi. 29. 5), 'cui scribendi magis est facilitas quam
facultas'.

[6] *E.* i. 11. 14; v. 17. 10; ix. 14. 6.

[7] *E.* iv. 8. 5.   [8] *E.* iii. 12. 5.

he tells us, in great haste,[1] and, though uninspired, they
are tolerable.

'My Muse,' said Sidonius, 'is sterile;'[2] and in saying
this he uttered the true and complete criticism of his
poetical works.[3] As we have seen, the great indictment
of the intellectual life of the fifth century may be based
on the fact that it was too easy, that is lacked stress and
strain. This indictment touches Sidonius as a writer
very closely. He wrote too easily for it to be necessary
for him to trouble much about the thought behind what
he was writing; he never seems to wrestle with an
idea. His subject-matter was the pagan mythology, and
though he could give a certain languorous grace to his
pagan *epithalamia*, he could not save them from insi-
pidity.[4] His epigrams are neat, but they lack the wit
to give them pungency. On the one occasion when he
deserts the pagan mythology to write about Christianity and its prophetic forerunners, one sees clearly
how he fails when he has not the tradition of great
writers behind him. He seems to see the greatness of
his subject, but when a new outlook, a new posture of
mind is demanded, his hand loses its cunning, he falters
and endeavours to treat the Hebrew prophets as he had
treated the gods and heroes of ancient Greece. But the
lightness of touch and playfulness of manner which had
saved him there from egregious blunders cannot blend
with the sincerity of feeling in which he appoached the
subject of the Christian religion. His attempt to treat

----

[1] Cf. *E.* vii. 9. 4. Peter (p. 156) alleges that these references to hasty writing
are only the commonplaces of epistolography. So they may be, and they may
be true all the same.

[2] *C.* ix. 319, cf. *C.* xiv. 25, 'nostram pauperiem'.

[3] On the poems of Sidonius, Fertig (iii, pp. 18–20) is very good.

[4] Cf. *C.* xi. 59–60, 'oscula sic matris carpens somnoque refusae / semisopora
leui scalpebat lumina penna'.

the prophet Jonah in the sprit of a hero of mythology is purely grotesque. 'The whale's belly,' he writes, 'resounds, as his food sings psalms within, and it retches with the weight of the dauntless prophet.'[1]

Sidonius does not often, it is true, touch the ludicrous in this manner: in general he keeps to his level of un-inspired craftsmanship. As a literary phenomenon such a man is no rarity; he is like the thousands of minor versifiers, who are spray on the sea of poetry through-out the ages. Yet to his own age this man was not a minor poet, he was respected as one of the greatest writers of his time. The explanation is not difficult to find. Sidonius lived in an age in which cultural develop-ment had not kept pace with progress in other fields. The culture was in fact in the age, but not of it. Still fumbling at the ideas of centuries before, it was trying with ever diminishing hopes of success to illustrate them with a new turn of phrase. Thus with a culture out of touch with the times, poetic criticism, as has already been shown, could base itself merely upon standards of form and then only within a conventional framework of metre. By these standards Sidonius was judged in his own day; and as far as we can see the judgement that gave his verses high praise was defen-sible in the light of the age. We are not entitled to judge him by any other standard to-day.

With the publication of this complete collection of his poems Sidonius entered on a period of literary silence which lasted for many years:[2] the political designs of Euric had now brought Auvergne into the

---

[1] *C.* xvi. 26–30, 'Resonant dum uiscera monstri / introrsum psallente cibo uel pondera uentris / ieiuni plenique tamen uate intemerato / ructat cruda fames, quem singultantibus extis / esuriens uomuit suspenso belua morsu'.

[2] *E.* ix. 12. 2.

vortex of events, and the next years of Sidonius' life were to be occupied with sterner duties than elegant writing.

The state of affairs was, indeed, such as to bring little comfort. Euric was now in unconcealed antagonism to the empire. In 469,[1] following exactly the advice of Arvandus, he appeared on the Loire with an army; Riothamus, who seems to have been taken unprepared, was attacked at Bourg-de-Déols, near Châteauroux, before the troops of Syagrius could come to his assistance. He was completely defeated and escaped with difficulty into Burgundian territory.[2] Though Euric did not as a result of this victory attempt to gain territory to the north of the Loire,[3] the whole of Berry fell into his hands.[4] With Berry and the greater part of *Narbonensis prima* in his possession, Euric could now attack from three sides Auvergne and such portions of *Aquitanica prima* as were still left to the empire. The situation would have been difficult, even if the authorities had been enthusiastic in defence.

But they were not: Magnus Felix,[5] the praetorian prefect, was, indeed, a loyal subject of the empire, but his subordinate, Seronatus,[6] the Vicar of *Aquitanica*, was an adherent of the Visigothic policy which Arvandus

[1] The date of the outbreak of war is fixed by Schmidt (i, p. 262, n. 2) from John Ant., fr. 206, 2 (*F.H.G.*, iv, p. 617).

[2] Jord., *Get.*, xlv. 238; Greg. Tur., *Hist. Franc.*, ii. 13 (18); cf. *E.* iii. 9. The 'Romani' of Jordanes (l.c.) can only refer to Syagrius or one of his generals. For the site of Dolensis Vicus see Longnon (*Géographie*, p. 466).

[3] See Longnon, pp. 41–2; Schmidt, i, p. 262.

[4] Greg. Tur., l.c. Some authors, such as Chaix (ii, p. 130), have recklessly disregarded this evidence, and refused to admit a Visigothic conquest of Berry until 474.

[5] *E.* ii. 3. 1. Cf. Gennadius, *De Vir. Ill.* 8. For the chronology of praetorian prefects at this time see Appendix D.

[6] On Seronatus see Yver, pp. 23–4; for his title, Stein, p. 580, n. 3. Earlier writers usually made him praetorian prefect.

had pursued. Like him, he was accused both of peculation and of treason. Not content, however, with giving treasonable advice, he was actually endeavouring to bring certain portions of his diocese under Visigothic rule, substituting, in fact, as Sidonius says, the laws of Theodoric for those of Theodosius.[1] As a result of such cessions of Roman territory, estates which had previously been on Roman soil became liable to the barbarian 'tertiatio', and their owners might suddenly find a Visigoth quartered on them.[2] With his continual visits to the Visigothic court,[3] the Roman governor seemed to be taking orders from Euric. Sidonius wrote to his brother-in-law Ecdicius in tones of the deepest despair. 'If the state cannot help us,' he declared, 'if, as rumour has it, the emperor Anthemius has no resource, the nobles are resolved to follow your lead, and abandon either their country or the hair of their heads.'[4] Only by accepting the tonsure of clerical office could a Gaul, he thought, remain faithful to the Roman tradition.

These words are in such a remarkable manner prophetic, that unless we believe that the prophecy was made up at a later date to justify the event, we can only feel that Sidonius had already, as he wrote them, an inkling of what his future might be.[5] In 469 (after July)

---

[1] E. ii. 1. 3. On the laws of Theodoric see Zeumer in *M.G.H.*, Leges, i, p. xiii.

[2] E. ii. 1. 3, 'implet cotidie siluas fugientibus uillas *hospitibus*'. 'Hostibus'— adopted by Lütjohann and Mohr from the better MSS.—gives a far less forceful reading and considerably weakens the parallel with 'fugientibus'. 'Hospitibus', the technical term for barbarians thus billeted on Roman estates, is the *difficilior lectio* and should be adopted.

[3] Seronatus returning from Toulouse, see *E.* v. 13. 1; from Aturres, *E.* ii. 1. 1. On Aturres (Atura) as an occasional residence of the Visigothic Court see *Praef. ad Leg. Rom. Vis.* (ed. Haenel, p. 4); for its geographical position, Longnon, pp. 595–6.   [4] *E.* ii. 1. 3. See Sirmond, p. 25.

[5] So Kaufmann, in *Neues schweizerisches Museum*, 1865, p. 10.

Q

Eparchius, bishop of Clermont, died and Sidonius was elected to occupy his place.[1]

That an aristocrat, who had pursued a successful career in the civil service, should thus suddenly be transformed into a bishop in spite of his lack of theological knowledge was less surprising then than it appears to us now. Many of the Gallic bishops were, indeed, distinguished theologians who had been educated at church schools and had already passed through the inferior grades of deacon and presbyter; many had been called from the monasteries, notably that of Lérins, to fill an episcopal vacancy. But when the bishop began to be not only the spiritual head of his flock but in some degree the civil administrator of his diocese, it was often considered advantageous to select a man of aristocratic birth, who had already some experience of the world. Thus Germanus had been a soldier, and Ambrose a distinguished civil servant, when they were called to be bishops. When Sidonius himself was asked to nominate a bishop for the see of Bourges, he named a layman who had previously been a *comes*,[2] and in his speech to the congregation he put the case in favour of appointing a man of the world to the episcopate. 'If I name a monk to you,' he said, ' I shall be told that such a man is better fitted to intercede for men's souls before the celestial judge than for their bodies before the judge of the world.' Such rapid ordinations as these elections involved were indeed technically illegal by canon law,[3] and in their pastoral

---

[1] Greg. Tur., *Hist. Franc.*, ii. 15 (21). For the date, which is that suggested by Mommsen (ap. Lütjohann, p. xlviii), see Appendix, pp. 205–7.

[2] *E.* vii. 9. 9.

[3] See *Nicaea*, 2; *Serdica*, 10; *Laodicea*, 3; and Bright, pp. 7–10. Cf. Turner in *C.M.H.*, i, p. 152; Löning, pp. 127–9.

letters popes of Rome had endeavoured to forbid them.[1]
But circumstances were too strong; sometimes candidates were, like St. Ambrose, rushed quickly through
the lower ecclesiastical degrees; but often this formality
was omitted, and the layman became at once a bishop.

The times indeed were such that an experienced man
of the world might make a very good occupant of a
bishopric. With the decline of the municipal authorities, the bishop became a representative of the people
against the oppression of the imperial governor, and an
emperor might listen to the complaints of a bishop
when he would not listen to any one else. Thus we
find St. Martin protecting members of his diocese
against the exactions of the count Avitianus,[2] and St.
Germanus undertaking a journey to Arles, the seat of
the prefecture, in order to obtain a remission of taxes
for his diocesans.[3] When the imperial government
grows weaker and cannot protect the towns from the
barbarians, it is the bishop who mediates for the
inhabitants,[4] who directs the defence,[5] and indeed
sometimes leads the inhabitants to battle.[6] Thus as
the power of the state declines the Church is found
gradually taking over its functions of government.

Such functions are never indeed formally legalized
by the state; they depend rather on the great moral
prestige of the bishop. Nevertheless, the state was quite

[1] Innocent, *Ep.* iv. 5; Celestine, *Ep.* ii. 3; Leo, *Ep.* xii. 4.

[2] Sulp. Severus, *Dial.* ii (iii). 4. Cf. Augustin, *Ep.* 153, and *Cod. Theod.*,
ix. 40. 24, where prisoners are released at the request of bishops.

[3] Constantius, *Vita S. Germani*, 21.

[4] For a light on the independent action taken by a bishop in the face of
barbarians, see the interesting story of the bishop of Margus in Priscus, fr. 2
(*F.H.G.*, iv, pp. 72–3).

[5] Eugippius, *Vita Seuerini*, ii. In ib. iv. 1–2, Severinus is seen issuing orders
to a tribune in command of soldiers.

[6] Constantius, *Vita S. Germani*, 18.

ready to use this prestige in its own service. The bishop
not only intercedes with barbarians on his own initiative; he may be requested to make a convention with
a barbarian king either by a provincial council,[1] or by
the emperor himself.[2]

Moreover, the state was quite prepared to make use
of the bishop as a powerful influence in the life of the
town. In selecting men to fill the office of *defensor
ciuitatis*, that curious device which the government had
created for a safeguard against its own exorbitances,
it is laid down by law that the bishop and clergy shall
take part;[3] thus the position of the ecclesiastical authority as defender against the exactions of officials is
recognized by the state, even if it does not legalize the
actions of bishops who, when the office of *defensor* had
fallen into decay, practically took over its functions
themselves.[4] To visit the prisoner and captive was a
duty imposed upon the bishop by the doctrines of
his faith; the state recognized and gave force to this
by compelling the magistrates to treat prisoners with
humanity and to permit the attendance of clergy at the
prisons.[5] The state was also prepared to assist by legislation bishops who were endeavouring to repatriate
captives taken by the barbarians.[6]

Another important sphere, in which the state was
prepared to use the assistance of the bishop, was that

---

[1] As Epiphanius in 475, see Ennodius, *Vita*, 81.

[2] Cf. Priscus, fr. 24, 1 (*F.H.G.*, iv, p. 102); Socrates, vii. 8. The peace of
475 was entirely transacted by ecclesiastics, *E.* vii. 6. 10; vii. 7. 4.

[3] *Cod. Iust.*, i. 55. 8.

[4] The opinion that the bishop in some instances actually was the *defensor
ciuitatis* cannot be sustained. (See Pfister, in Lavisse, ii, p. 24.)

[5] *Cod. Theod.*, ix. 3. 7.

[6] On the redemption of captives by bishops see Eugippius, *Vita Seuerini*, ix.
1–3; Ennodius, *Vita Epiphanii*, 141 *et seq.*; Socrates, vii. 21; *Hist. Misc.*, xiv. 18.
For legislation by the state see *Cod. Theod.*, v. 5. 2; *Const. Sirmond.*, xvi.

of civil justice. Civil lawsuits were supposed to be tried by the provincial governor, but we can well believe that the press of business must have been so great that even the most upright judge was tempted to override the strict letter of the law, if only to get a lawsuit finished. In the later empire, Roman justice was proverbially bad and uncertain,[1] probably because the judiciary was under-staffed as much as for any other reason. To remedy this, Julian attempted to revive the *pedanei iudices*, or judges who went on circuit in petty cases.[2] The idea was good, but little seems to have come of it, for there are no later mentions of *pedanei iudices* in the Codes. More fruitful was the idea of enlisting the bishops to help reduce the press of litigation. Before the recognition of Christianity, bishops had been the dispensers of justice within the Christian community;[3] and when it was established as the state religion emperors were quite ready to take advantage of this already existing institution.[4] Episcopal justice was expeditious, and the judges were free from the blight of corruption which poisoned the imperial civil service.[5] As early as 318, Constantine had allowed a litigant to transfer his case to an episcopal court, even if the trial before a civil judge had already begun;[6] and in 333 this permission was still more clearly expressed: an action could be transferred by either litigant to an ecclesiastical court even if the other party did not wish it.[7]

---

[1] *Cod. Theod.*, i. 16. 7; Amm. Marc., xxx. 4. 9-19.

[2] *C.I.L.*, iii. 459; iii. 14198; *Cod. Theod.*, i. 16. 8. Cf. *Cod. Iust.*, iii. 3. 1-5.

[3] For the origins of this jurisdiction see Lardé (pp. 17-20), and in general Löning, pp. 229-30.

[4] Cf. Eusebius, *Vit. Const.*, i. 42; iv. 27. 2; Sozomen, i. 9.

[5] See Vinogradoff in *C.M.H.*, i, pp. 565-6.

[6] *Cod. Theod.*, i. 27. 1.

[7] *Const. Sirmond.*, i. On the relation between this law and the preceding,

Subsequent legislation[1] to some extent modified these extreme privileges of the ecclesiastical courts, and a law of Valentinian III[2] limited their power to actions in which both parties consented to the arbitration of the bishop.

Nevertheless, there is good evidence that these ecclesiastical tribunals were extremely popular. Ambrose declares that the last hours of his day are occupied with the examination of cases which had been removed from the prefect's court[3]: the time which Augustine would gladly have devoted to other uses had to be occupied in listening to the noisy wrangling of disputants. 'I feel that I should be allowed to address them,' he writes, 'in the words of the psalmist: "Depart from me, all ye that work iniquity"'[4]; Chrysostom complains that he has more legal questions to unravel than a civil judge,[5] and the position of the bishops as interpreters of the law is illustrated by the fact that they are included among the authorities consulted by Alaric before the publication of the *Lex Romana Visigothorum*.[6]

In the correspondence of Sidonius there is at least one clear picture of the operations of an episcopal court. It is to be found in a letter addressed to Lupus, bishop of Troyes.[7] A woman had been abducted by brigands and sold into slavery at Troyes, where she had subsequently died.[8] Her kinsmen bring an action against the

see Seeck, *Regesten*, pp. 7–8. Godefroy's opinion (ad *Cod. Theod.*, i. 16) that *Const. Sirmond.* i is a medieval forgery is now generally abandoned. (See Maassen, *Geschichte der Quellen und der Litteratur des kanonischen Rechts*, i, pp. 792–6.)

[1] On the progress of this legislation, see Lardé, pp. 69–80.

[2] *Novell. Valentin.*, xxxiv.   [3] Ambrose, *Ep.* 49

[4] Augustin, *de Opere Monachorum*, xxxvii; and *Serm.* 24 ad Psalm. cxviii.

[5] Chrysostom, *De Sacerdotio*, iii. 17.

[6] *Praef. ad Leg. Rom. Vis.* (ed. Haenel, p. 2).   [7] *E.* vi. 4.

[8] Ib. § 2, 'in domo defungitur'. Dalton's translation (ii, p. 84), 'is under the roof of', must surely be incorrect.

purchasers, and Sidonius writes to the bishop urging him
to make such a decision as will satisfy both parties. The
case has considerable legal interest.[1] According to the
law, those who dealt with a kidnapped woman could
be punished with death,[2] but as this was a capital
charge it could be compounded, and the accuser was
permitted to drop the action[3] and accept a sum of
money in 'composition'. That sum the bishop is asked
to fix. We see clearly in this case an approximation to
the Teutonic law of 'composition fines', the case also
shows how the practice of 'composition' gives the
bishop in a rather circuitous manner what the law had
denied him—criminal jurisdiction, at least in capital
offences.[4] The language of the letter seems to show
that the parties approach a regularly constituted court,[5]
and that a regular judicial procedure is contemplated;
though whether in these times, when the civil government had decayed, the bishop could have recourse to
any power in order to enforce his judgement must
remain uncertain. Probably, as Sidonius seems to imply, it was the authority and prestige of his sacred
office that gave validity to his decisions. No other letter
of Sidonius refers so definitely to proceedings taken by
a layman before an episcopal court. On one occasion
Leontius, bishop of Arles, is requested to use his influence for a litigant who wishes to obtain good legal advice.[6] More interesting is the letter[7] in which Sidonius

---

[1] It is discussed by Esmein in *Revue générale du Droit*, 1885, p. 305 et seq.

[2] *Cod. Theod.*, ix. 24. 1. [3] *Cod. Iust.*, ii. 4. 18.

[4] For 'compositio' cf. *E.* v. 19. 2 and *E.* v. 13. 4. A very close parallel to this
case is seen in *Formulae Turonicae*, 16 (quoted by Esmein, l.c., p. 307).

[5] Cf. 'Iudicium', *E.* vi. 4. 3.

[6] *E.* vi. 3. 2.

[7] *E.* vi. 2. It is doubtful whether the words 'quibus in eos *nouum ius* professio
uetustumque faciebant amicitiae' (§ 3) contain any reference to a court.

examines the dispute between Eutropia, a venerable
matron, and Agrippinus, a cleric. He tells us how
he has used the old right given him by friendship and
the new right of his position as bishop in his efforts to
bring about a reconciliation. 'The dispute,' he writes
to a neighbouring bishop, Pragmatius, 'is now to be
brought before you'. Sidonius urges him to give a
decision which will ensure harmony between the parties; Eutropia, he says, would welcome even an adverse
decision if it saved her from ligitation. This ligitation
seems to be distinguished by Sidonius from proceedings
in the ecclesiastical court of the bishop, and it appears
that in this case Pragmatius acts not as a judge but as
an unofficial meditator, who will save the parties from
the necessity of litigating, presumably in a civil court.[1]
There is indeed more than one mention[2] in Sidonius of
laymen who conduct such arbitrations. The evidence
from the letters, then, would seem to show that though
in some sees the bishop did indeed sit as a judge of civil
matters in a court and give officially binding decisions,
in general the bishop is rather an unofficial mediator,
whose decisions merely operate by the goodwill of the
disputants and are intended to save them from the evils
of litigation.

Among his own clergy, however, the bishop's authority was paramount,[3] and it extended over the monasteries in his diocese.[4] A priest could not even leave the
diocese to which he belonged without obtaining a letter

---

[1] If the rule laid down in the Council of Angers was upheld (*Conc. Andec.*
(453), 1; cf. *Conc. Arelat.* (452), 31), Agrippinus would not in any case have been
allowed to litigate in a civil court without the bishop's leave.

[2] *E.* ii. 7. 1; iii. 5. 2; iv. 6. 4; v. 7. 2. Cf. Wisslack in *P.-W.*, ii, p. 411.

[3] *Cod. Theod.*, xvi. 2. 41, 47; *Conc. Arelat.* (452), 31; *Conc. Venet.* (463), 9;
Bright, pp. 155-8.

[4] Cf. *Conc. Aurel.* (511), 7, 49.

of authorization from the bishop.[1] In law-suits of
the clergy the bishop's decision was final; appeal to the
civil governor was strictly forbidden.[2] A canon in a
council of slightly later date suggests that the absolute
dependence of the lower clergy on the bishop was further assured by the fact that it was he who paid their
salaries.[3]

This last fact suggests another reason why a trained
administrator was desirable as bishop in a diocese. The
Church was very rich. From an early date it had been
allowed to receive legacies or donations;[4] and though
in the East a bishop might impoverish not only himself
but his church by charitable expenditure,[5] in the West
since the council of Carthage in 398 alienation of property was forbidden except under circumstances of
extreme necessity,[6] and by the rules of the later council
of Agathe in 511 it was only allowed if the bishops of
three provinces in the diocese consented to it.[7] Donations to the Church indeed were frequent; sometimes
the bishop himself bequeathed property to it on his
death,[8] and there was a continual stream of donations
from private persons. A letter from Sidonius praises a
friend for presenting a farm to the church of Clermont,[9]
and in another he urges donations to the Church, 'for,'
he says, 'whatever you give to the church is really
gathered in for yourself.'[10] Thus the Church was continually acquiring lands, sometimes at a considerable

---

[1] *Conc. Turon.* (461), 12; *E.* vi. 8. 2; vii. 2. 1; ix. 10. See Sirmond, pp. 66–8.
[2] *Conc. Andec.* (453), 1.
[3] *Conc. Agath.* (506), 63, 'clerici omnes . . . stipendia . . . a sacerdotibus consequantur'.
[4] *Cod. Theod.*, xvi. 2. 4; *Cod. Iust.*, i. 2. 1, 13, 14.
[5] Sozomen, i. 11.   [6] *Conc. Carth.* (398), 31, 32.
[7] *Conc. Agath.* (506), 7.   [8] Greg. Tur., *Hist. Franc.*, x. 31.   [9] *E.* iii. i. 3.
[10] *E.* viii. 4. 4; cf. *E.* iv. 11. 5.

distance from the city,[1] and though the lands themselves were subject to the usual land-tax[2] the bishops
and clergy were exempted from the compulsory 'mu-
nera', which pressed so heavily upon the ordinary
population.[3] In addition to its revenue from landed
property the Church was permitted to receive income
from the profits of a shop or factory,[4] and though
Councils strictly forbade a cleric to lend money upon
usury,[5] there is evidence that the rule was not strictly
observed.[6] We need not therefore be surprised that the
yearly income of a bishopric in the East might amount
to as much as two thousand pounds sterling,[7] and that
even in sixth-century Gaul a bishop could die leaving
more than twenty thousand pieces of gold.[8] When
Pope Gelasius shortly after this enacted that a careful
distinction should be made between the revenue of the
bishop and that of the Church,[9] it must have been an
action that expediency demanded. But until that distinction was drawn the wealth a of bishopric was a particular temptation for an unscrupulous man.

It is obvious, indeed, that election to the bishopric
entrusted a man with not only a position of great respon-
sibility in spiritual and temporal matters, but also with
the opportunity of administering property of ever-in-
creasing richness. There was a temptation for unworthy
men to put themselves forward as candidates merely in

[1] Greg. Tur., *Hist. Franc.*, ii. 26 (36).   [2] Löning, pp. 228–30.

[3] *Cod. Theod.*, xvi. 2. 40; *Const. Sirmond.*, xi.

[4] *Cod. Theod.*, xvi. 2. 10.

[5] *Conc. Nic.*, 17; *Conc. Carth.* (398), 16. See Bright, pp. 56–9.

[6] See the curious incident in *E.* iv. 24. 6, with Dalton's note (ii, p. 234).

[7] *Nov.*, cxxiii. 3.

[8] Greg. Tur., *Hist. Franc.*, x. 31. Compare the story of Patiens, bishop of
Lyons, who is alleged to have provided food for four thousand persons (Greg.
Tur., *Hist. Franc.*, ii. 16. (24); cf. *E.* vi. 12. 5–9).

[9] See Jaffé, *Concilia*, iv, p. 1195.

order to gain the power to plunder the Church estates,[1] and such men would frequently disburse huge sums in an attempt to secure their election.[2] To prevent such scandals as far as possible, the conditions of episcopal elections were very carefully controlled.

Originally, as it seems, the bishops were elected by their congregations without any interference from the other bishops of the province.[3] It was necessary, however, that the person elected should be consecrated by at least one other bishop, and thus the other provincial bishops came in practice to have some control over the choice of the congregations. The bishops had, indeed, the final word, because in the last resort they could always refuse to consecrate a man whose name was submitted to them; but the exact relation between the congregations and the provincial bishops was never fixed, indeed it seems to have varied between one part of the empire and another; and the history of episcopal election in the fourth and fifth centuries is that of a continually fluctuating struggle between the congregations and the bishops. At the Council of Laodicea in 341 it was laid down that the multitude should have no part in the election of a bishop;[4] and the letters of Celestine and Leo to the bishops of south Gaul,[5] which forbid the appointment of bishops against the will of the populace, can only imply that there at least no sort of popular election existed. A law of Valentinian III,[6] indeed,

---

[1] *E.* iv. 25. 2. For a picture of the worldliness of Gallic bishops see Sulp. Severus, *Chron.*, i. 23. 5; *Dial.*, i. 21. 3–4.

[2] *Edictum Glycerii*, in Haenel, *Corpus Legum*, p. 260; cf. *E.* vii. 5. 2.

[3] For the early history of episcopal elections see Turner in *C.M.H.*, i, pp. 152–4.

[4] *Conc. Laod.* 13.

[5] See Celestine, *Ep.* IV. v. 7 (Migne, *Patr. Lat.*, l, p. 431); Leo, *Ep.* x. 4 (Migne, *Patr. Lat.*, liv, p. 632).

[6] *Novell. Valentin.*, xvi (A.D. 445); cf. § 1, 'inuitis et repugnantibus'; Leo, *Ep.* x. 6,

expressly blames the metropolitan of Arles for consecrating bishops to whom the inhabitants take
objection; it even appears that sometimes the newly
appointed bishop could only enter the town under the
protection of an armed guard. Such procedure is quite
inconsistent with anything like a popular election.

On the other hand there were sees where the tradition of popular election by the clergy, nobility, and
commons was very strong. Martin was elected to the
bishopric of Tours by popular acclamation in spite of
the opposition of certain bishops,[1] and Germanus of
Auxerre, according to his biographer, was elected by
the unanimous consent of clergy, nobility, and commons, not a word being said about the provincial
bishops.[2] It is doubtful whether we can put much
weight on the accounts in Gregory of Tours of episcopal elections held many years before his time; yet it is
worth noting that in his account of the election of
Brice at Orléans popular choice is clearly implied.[3]

Nevertheless, to free popular election there were obvious objections; the number of electors was frequently
very large, and thus the proceedings were tumultuous
and the results not always desirable. On the other
hand, the people did not wish to forfeit entirely their
rights to the election of the bishop. It was natural
therefore that compromises of different kinds were
evolved between the extremes of free popular election
and nomination by the other provincial bishops.

In Italy, especially, the local clergy act as interme-
diaries between the populace and the bishops; they
either approve the candidate put forward by the laity,

---

[1] Sulp. Severus, *Vita S. Martini*, 9. 3.   [2] Constantius, *Vita S. Germani*, 11.
[3] Greg. Tur., *Hist. Franc.*, ii. 1.

and offer him to the bishops for consecration, or they themselves carry out the election, and submit the name of their nominee to the people for commendation.[1] This derogation of authority to the clergy seems, however, not to have existed in Gaul until a much later date.[2]

In Gaul the efforts of Celestine and Leo,[3] after the scandalous affairs in the diocese of Arles, had succeeded in establishing the rule that the bishops should preside over a free election of the people. The latter recommend a man to the bishops, who in consecrating him, says Leo, must be sure that they are acting on the wishes of clergy, nobles, and commons.[4] It is clearly implied, however, that the bishops could refuse to consecrate an obviously undesirable candidate, and thus in spite of the popular election the ultimate decision lay with them. In 452, however, a compromise of quite a different nature was framed by the Council of Arles. Three candidates were to be put forward by the bishops, and one of them was to be chosen by the congregation.[5] Considered as a device for removing the opportunity for tumults at elections, this compromise cannot be considered as very happy.

It is unfortunate that we have no information about the details of Sidonius' own election: we possess,

---

[1] Turner in *C.M.H.*, p. 153.

[2] On episcopal elections under the Merovingian kings see Dalton, *Gregory of Tours*, i, pp. 288–300. It is noteworthy that the Formula, which is addressed to the king requesting his confirmation of an episcopal appointment, comes in Merovingian times from 'omnes ciues' (*Form. Marculf.*, i. 7): in the similar Formula used in the Carolingian age, the bishop is no longer elected 'consensu omnium ciuium', but 'consensu totius cleri'. (*Form. Marculf. Aeu. Carol.*, 12.)

[3] See letters of Celestine and Leo cited above. An election conducted under this rule is described in Greg. Tur., *Hist. Franc.*, ii. 12 (13).

[4] Leo, *Ep.* x. 6 (Migne, *Patr. Lat.*, liv, p. 634); cf. *Ep.* xl (ib., p. 815).

[5] *Conc. Arelat.* (452), 54.

however, two letters from his hand, in which he describes
the election of other bishops. In the election of Châlon[1]
he was only an eye-witness, and was indeed a layman at
the time, but in the other, at Bourges in 470,[2] it was
Sidonius who actually nominated the bishop. In both
accounts there are difficulties, and in neither election
does the procedure seem to have been quite normal,
yet we have no other documents which throw so much
light upon the methods of election at this period.

The account of the election at Chalon may be sum-
marized as follows: the metropolitan, Patiens, bishop
of Lyons, accompanied by several of his provincial
colleagues (the number is not stated), goes to the town
in order to consecrate a bishop. He finds that the
sentiments of the townspeople are much divided;
there are three candidates, none of them very suitable.
The bishops meet in council, and at the instigation of
Patiens and Euphronius, bishop of Autun, they determine to reject all three candidates and to elect an
entirely different person, who had not been nominated
at all. This was not to the liking of some of the citizens,
but it is clearly implied that the procedure is perfectly
legal. We see at once that this election is not decided
according to the rule so recently laid down at Arles;
yet no later council contains a specific revocation of
that rule, which must simply have lapsed. On the
other hand, in the record of this election, we again find
the mention of three candidates. It is tempting to
assert that this identity of number is no mere coincidence. And this assertion gains plausibility when we
recollect that these are not the only passages in which

---

[1] *E.* iv. 25. Löning (pp. 118–20) is superficial.
[2] *E.* vii. 9. Cf. *E.* vii. 8; vii. 5.

the number three occurs in reference to episcopal elections. A rule established in the East by a law of Justinian in 528 lays down that in the event of an episcopal election the congregation shall put forward three men from whom the metropolitan and other bishops shall select their colleague.[1] It is to be noted that this is, with one exception, precisely the procedure adopted at Chalon. The exception is that the bishops rejected all the popular nominees, and bestowed the bishopric on an entirely different man. Seeing that the popular election had resulted in candidates appearing who were not truly representative, the bishops followed the advice contained in the letter of Leo, and nominated of their own accord a bishop who seemed really satisfactory to the congregation. The proceeding was unusual and indeed evoked some surprise,[2] but it was not illegal any more than the violent ordinations of bishops by the metropolitan of Arles were strictly speaking illegal. We see then that the procedure adopted at Chalon inverts, as it were, that laid down at Arles, and follows that subsequently adopted in Justinian's code. That the law is some fifty years later need not surprise us, for on such a point it is quite to be expected that practice should outstrip imperial legislation.

It would appear at first sight that there is no analogy between the election at Bourges and that of Chalon, yet on attentive examination we find that the principle is the same. According to the data given by Sidonius, there are numerous factions and the candidates occupy more than two benches.[3] 'If the people had not grown

---

[1] *Cod. Iust.*, i. 3. 41 (42). Cf. *Novell.*, cxxiii. 1; cxxxvii. 2.
[2] 'stupentibus factiosis erubescentibus malis, acclamantibus bonis reclamantibus nullis collegam sibi consecrauere'. *E.* iv. 25. 4; cf. ib. § 1, 'illa quae bonum publicum semper euertunt studia priuata'.   [3] *E.* vii. 9. 2.

reasonable,' he writes, 'and submitted their judgement to that of the bishops, there would have been little chance of effecting anything.' This certainly seems a clear implication that there might, at least, have been a popular election. On the other hand, in his letter to Agroecius of Sens in which he invites his co-operation in the election, Sidonius writes as follows: 'I have done nothing to prejudice the right of decision which belongs to your office',[1] and to another bishop, Euphronius, he writes promising that whatever instructions he may give, 'will be followed by the bishops and applauded by the people'.[2] Here on the other hand we seem to have equally clear evidence that the election was conducted by bishops. How can the discrepancy be explained? It is to be explained by realizing that it does not properly exist. If the election at Chalon had proceeded normally, according to the rule of Justinian, there would have been an election in two stages, one of them conducted by the congregation, the other by the clergy. At Bourges the normal procedure would have been adopted, but it broke down in a different way; the people were unable to conduct the preliminary election at all owing to the violence of the factions, and Sidonius was therefore requested to nominate without being assisted, as it were, by a preliminary weeding-out. And the discrepancy between the three candidates at Chalon and the two benches of candidates at Bourges is again to be explained by realizing that it is not properly a discrepancy. The three candidates at Chalon are those who have survived the popular election, the two benches of candidates at Bourges are those who have yet to undergo it. At Bourges, therefore, Sidonius

<hr>

[1] *E.* vii. 5. 4.   [2] *E.* vii. 8. 4.

proposes that the preliminary popular election should be omitted and that he himself shall choose the bishop from the mass of the candidates. The people agree, and Sidonius makes them swear an oath to abide by his decision. He then chooses, not, as at Chalon, a completely different person, but one of the candidates,[1] Simplicius. Thus it is seen that both these elections are abnormal examples of the same type, and that both followed, not the rule of the Council of Arles, but a rule subsequently affirmed by imperial rescript in the East.[2]

[1] That Simplicius was a candidate is clearly proved from *E.* vii. 8. 3.

[2] For convenience a table is subjoined explaining this theory of the elections. Steps which are conjectural are enclosed in brackets.

$J =$ the 'Justinian rule'.
$C =$ the procedure at Chalon.
$B =$ the procedure at Bourges.

J. Many candidates appear—from whom three are chosen by popular election—of these three the bishops select one.

C. (Many candidates appear from whom) three (are chosen by popular election), and of these three the bishops—exceptionally—refuse to select one, but substitute a nominee of their own.

B. Many candidates appear (from whom three) would be chosen by popular election—but, as the people cannot decide, the bishop with their consent (allows the popular election of three candidates to be omitted) and himself selects one out of the (original) candidates.

VII

SIDONIUS THE BISHOP AND THE SIEGES
OF CLERMONT

IT has been supposed by some that Sidonius was ambitious for the honour of election to the see of Clermont
because an ecclesiastical career offered him greater opportunities for power.[1] While it may, as we have seen,
be a correct conjecture that he contemplated the possibility of entering the Chuch, there are passages which
seem to indicate that this election actually came as a
surprise to him and was even contrary to his wishes.[2]
When he was elected, he felt at once that he was unworthy to watch over the spiritual destinies of his flock;
the mental strain of the sudden change in his manner
of living brought on a fever,[3] and to the end of his life
he was never tired of calling to mind his unworthiness
of the position to which he had been called. When the
bishops of the neighbouring dioceses wrote to congratulate him upon his new elevation,[4] his deference was
so extreme as almost to appear unnatural. In his letter
to Lupus, bishop of Troyes, he declares that he is 'a vile
clod of earth, foul with sin', that the recollection of his

---

[1] Ampère, *Hist. littéraire de la France*, ii, p. 247.

[2] *E.* iv. 3. 9, 'impactae professionis obtentu'; v. 3. 3, 'tantae professionis
pondus impactum'; vi. 1. 5, 'oneris impositi massa'; vi. 7. 1, 'indignissimo mihi
impositum sacerdotalis nomen officii.' These passages take much of the sting
from the argument of Kaufmann (*Neues schweizerisches Museum* (1865), p. 19),
who asks why, if Sidonius felt his unworthiness to such an extent, did he consent
to become bishop at all.

[3] *E.* v. 3. 3. Whether this illness is the same as that described in *E.* ix. 14. 1
it is impossible to say.

[4] The letter of Lupus congratulating Sidonius on his election, of which earlier
biographers made use, has been proved to be a forgery of the seventeenth
century. (See J. Havet in *Bibl. de l'École des Chartes*, xlvi, pp. 252–4.)

past life has become a misery to him; he is a 'despicable worm', to be compared to the Gerasenes and the lepers of the scriptures.[1] 'How am I', he writes, 'to intercede for the sins of my people, when I need intercession so much myself.' His conscience was covered with gaping wounds, and needed to be washed clean with tears.[2] Throughout his career he employed the same language to priests and even to laymen;[3] as he grew older, he became merely 'a veteran in transgression',[4] and when he awaited death, he could only hope to appear before the judgement-seat as a servitor, a hewer of wood and drawer of water for his betters.[5]

That these professions had an element of sincerity in spite of their turgid language we need not doubt; they are symptoms of the rapt enthusiasm with which Sidonius approached and was absorbed in his new career, an enthusiasm which is a foundation of his character. Moreover, they had some justification. As a layman, Sidonius had been a good man by conventional standards, and his piety, though expressing itself for the most part in mechanical prayers for the aid of God before an undertaking,[6] had been real and sincere. He had taken an intelligent interest in the election of a bishop in the neighbouring see of Chalon,[7] and was always ready if asked to provide an inscription for a church.[8] He had taken part in Church festivals in a reverent spirit, though one may doubt whether an earnest believer would have used the interval between

---

[1] *E.* vi. 1. 2, 4, 5.   [2] *E.* vii. 6. 3.   [3] *E.* v. 3. 4.
[4] *E.* ix. 2. 3.   [5] *E.* ix. 8. 2; cf. ix. 3. 4.
[6] See *E.* i. 6. 1; i. 9. 8; ii. 2. 3; ii. 12. 3; ii. 4. 2; iii. 6. 3; vii. 15. Though it is true that the art-form which Sidonius used in his earlier books did not encourage devotional sentiments (cf. p. 172), the pagan spirit of the epitaph to Filimatia (*E.* ii. 8. 3.) lends weight to the inference drawn from these passages.
[7] *E.* iv. 25.   [8] *E.* ii. 10. 4; iv. 18. 5.

early service and tierce to take part in ball-games and
dicing.[1] But his theological knowledge does not seem
to have been large; the poem to Bishop Faustus, as has
been shown, treats the sages of the Old Testament in
the manner of the heroes of Greek mythology, and
when the author touches upon the subject of the Incar-
nation he employs, without apparently realizing it,
language which can be construed as heretical.[2] We
need not be surprised that, when he became a bishop,
in more than one letter he asked his colleagues for
instruction in doctrine;[3] it was what he needed.

There is another lesson to be learned from the poem
to Faustus: from it we see that Sidonius seems much
better acquainted with the Old than with the New
Testament. A poem describing the sages of the Church
which mentions Elijah, Elisha, Judith, and Jonah and
has nothing to say of a single Apostle is not a little surprising. Even as a bishop he seems to have known his
Old Testament better. One of his rare quotations from
the New is a passage from St. Luke introduced to prove
the value of aristocratic birth;[4] whereas from the Old
he makes mention of Moses, Hagar, Sarah, and Abraham,[5] and in a letter to Bishop Principius of Soissons
he shows a curiously minute knowledge of the Mosaic
Law.[6] One may believe that this Gallo-Roman aristocrat, imbued with the spirit of the pagan mythology,
who could understand a personality better than an
idea, was more attracted by the Old than by the New

[1] *E.* v. 17. 3, et seq.

[2] *C.* xvi. 40–2 '(deus) . . . quique etiam nascens ex uirgine semine nullo.
/ ante ullum tempus deus atque in tempore Christus, / ad corpus quantum
spectat, tu te ipse creasti'. See Arnold in Hauck's *Realencyklopädie,* xviii, p. 308.

[3] *E.* vi. 3. 1; vi. 6. 2.   [4] *E.* vii. 9. 17.

[5] *E.* vii. 17. 2, ver. 28; viii. 13. 4; cf. vii. 6. 4; vii. 9. 21.

[6] *E.* viii. 14. 4–7.

Testament. It is not without significance that the only Paul mentioned in his works is not Paul the Apostle, but Paul an Egyptian ascetic.[1] When we find Sidonius supporting such a savagely retributive theory of punishment as leads him to say that the death of a murderer at least affords the satisfaction of revenge to the survivors of the murdered,[2] one begins to wonder whether he understood the lessons of the New Testament at all: such language becomes an old Roman rather than a Christian bishop.

Indeed, in the life of Sidonius the bishop reminiscences of Sidonius the layman are not infrequently to be found. A comparison between a Christian bishop and the hero of pagan mythology, Triptolemus, is introduced into a letter[3] without any feeling of incongruity. Less easy to pardon is the gross flattery which he poured on the heads of bishops. To one of them he writes that fame cannot sing all his praises;[4] Lupus of Troyes is the St. James of his day[5] and is congratulated on the incomparable merit of his Apostolic life.[6] Such examples are bad enough, but they are not the worst. Sidonius was still enslaved by the demoralizing influence of the rhetorical tradition. In a letter to Bishop Patiens of Lyons, who had by his bounty relieved a famine in south Gaul, Sidonius writes that it was worth while for the starving populations to lack food, if they could not otherwise experience his generosity.[7] This

---

[1] *E.* vii. 9. 9. Contrast this with the fact that in the letters of the thinker Faustus, which only occupy one fifth of the space, Paul the Apostle is mentioned twelve times.

[2] *E.* viii. 11. 13, 'nam quotiens homicida punitur, non est remedium sed solacium uindicari.'

[3] *E.* vi. 12. 6; Germain, p. 70, n. 6.

[4] *E.* viii. 13. 2.   [5] *E.* vi. 1. 1.

[6] *E.* vi. 4. 1.   [7] *E.* vi. 12. 9.

was falsehood and it was not new at that: sixteen years
before Sidonius had used almost the same words to the
emperor Majorian;[1] the subject had changed but the
panegyrist was the same. 'True praise adorns, false
praise is chastisement'; the words were Symmachus',
and Sidonius had quoted them with approval.[2] It is
unfortunate that he so easily forgot them.

Nevertheless, the parallel with the panegyric of Majorian has also a deeper significance. It was a sign of
the times, indicating perhaps a general conception as
much as the point of view of Sidonius himself. The
bishop had come to occupy the position as central
figure of panegyric that the emperor had once filled; it
is a testimony to the prestige of the episcopate to which
Sidonius himself in other passages bears witness. He
was not blind to the fact that he was greater as a bishop
than he had been as an official. When he is writing to
the ex-prefect Tonantius Ferreolus he does not speak
of himself as a Gibeonite and humble servitor in
heaven, but declares outright that in the opinion of
reasonable men the least of the clergy is beyond dispute above the holder of the highest office.[3]

Because he still remained a man of the world ignorant of the causes of theological anathemas, he was
an easy-going bishop. In one of his letters he writes
that a friend 'is perhaps none the worse for not being a
perfect paragon',[4] and such language, though it exempts

---

[1] 'quibus operae pretium fuit fieri famem suam periculo, si aliter esse non
poterat tua largitas experimento'. Cf. *C.* v. 585–6, 'fuimus uestri quia causa
triumphi, / ipsa ruina placet'. This very curious parallel was noted by Kaufmann (*Neues schweizerisches Museum*, 1865, p. 19).

[2] *E.* viii. 10. 1. This sentence is not to be found among the extant work of
Symmachus.

[3] *E.* vii. 12. 4; cf. *E.* iv. 14. 3.

[4] *E.* iv. 4. 3.

him from the charge of priggishness that laudatory biographers have tended to thrust upon him, must none
the less be felt as strange coming from the mouth of a
bishop. 'I am not,' he confesses frankly, 'of the stuff of
which martyrs are made: though I respect the austere,
I cannot like them.'[1] And as a bishop he preserved this
attitude of mind. In the controversies of Augustinians
and semi-Pelagians he maintained a neutral position;
his theological ignorance at least saved him from being
a fanatic. Claudianus Mamertus and Faustus of Riez,
the two protagonists, were both his friends. Claudianus
dedicated to him the *De Statu Animae*,[2] and to Faustus
he wrote that 'the heresiarch who challenged him
would be laid low'.[3] These words tell us clearly enough
that he did not understand the points at issue[4] and did
not see that Faustus' own opinions were of questionable
orthodoxy. It might indeed have surprised him to
learn that the man whom he celebrated as most blessed
above all in his generation would have his writings
condemned by the Church and be himself deprived of
the honour of canonization.[5]

Sidonius' own theological views, when we can discover them, are seen to be simple. He believed that
good or ill fortune in this world were part of a system of
divine rewards and punishments, and pushed this doctrine to somewhat strange lengths. Not only does he
ascribe the siege of Clermont to divine punishment for

[1] *E.* iv. 6. 2; vii. 4. 3.

[2] Claud. Mam., *De Statu Animae*, Praef., and i. 1; cf. *E.* iv. 2 and iv. 3.

[3] *E.* ix. 9. 15.

[4] Allard (p. 113) alleges that Sidonius asked Claudianus Mamertus to write
the *De Statu Animae* in order to refute the views of Faustus. Allard derives this
from his own imagination.

[5] Krusch, Preface to text of Faustus in *M.G.H.*, pp. lix–lx; Chaix, ii, p. 340,
n. 2.

an unknown sin,[1] but, when a correspondent had unexpectedly received an inheritance, Sidonius reminded
him that this was the direct intervention of God to
recompense him for a gift that he had made to the
Church.[2] In general, however, any one who searches
the works of Sidonius for theological views[3] will be
disappointed.[4]

Perhaps this lack of theological knowledge helps to
explain the fact that Sidonius never displays the feeling
of a persecutor. He even has a kindly word for the
Jews, and it is typical of this spirit of worldly tolerance,
which is perhaps more congenial to us than to contemporary ecclesiastics, that he points out how a Jew can
be defended as a man even if his creed be assailed.[5] In
these words the aristocrat gives a deserved lesson to the
Church of his day.

Of the manner in which he performed his duties in
the Church we have little information. If, as seems to
be implied in one of the letters, the public prayers were
interrupted by refreshments,[6] one can only consider that
his government of his flock was lax. As might be expected from a trained rhetorician, his preaching was

---

[1] *E.* iii. 4. 2; vii. 10. 1.   [2] *E.* iii. 1. 3.

[3] He is not without value, however, as a witness for the development of
Church discipline and institutions. There are two references in the letters (*E.* iv.
13. 4 and iv. 14. 3; cf. Ruricius, *Ep.* i. 8. 1.) to auricular confession, and *E.* vii. 6. 9
seems to imply that consecration is not valid unless the dying priest gives his
blessing to his successor.

[4] It is true that in Gennadius, *de Vir. Ill.*, 92, he is called 'doctor insignis'.
This passage is, however, only found in a few MSS. and is probably later than
the rest (see Mommsen, ap. Lütjohann, p. xliv). Sidonius wrote a volume called
'Contestatiunculae' (*E.* vii. 3. 1), which may be identical with the 'Missae' on
which Gregory of Tours wrote a commentary (*Hist. Franc.*, ii. 15 (22); Tillemont, *Mém. Ecclés.*, xvi, p. 277). Both the book and the commentary are lost.

[5] *E.* vi. 11. 1; iii. 4.

[6] 'erant quidem prius (sc. quam a.d. 473) . . . supplicationes, quae saepe
interpellantum prandiorum obicibus hebetabantur', *E.* v. 14. 2.

eloquent.[1] Later tradition alleged that he could speak without preparation on any subject that he chose:[2] once, it is said, when an enemy had removed the book which he used in performing the offices, Sidonius conducted the whole service from memory.[3] Such anecdotes, though unconfirmed, are at least consistent with the man as he is revealed to us in the letters.

As an administrator in his diocese Sidonius was true to the principles of his life: he tried to do his duty in his new station. Detractors accused him of the old vice of disdainful pride,[4] and there were others who insinuated that he was a vain man, the sincerity of whose faith was uncertain.[5] That he was in heart the old student of rhetoric, the old worshipper of ancient heroes, we may well believe: that he was considered a vain man is a valuable corrective of the idea that one would gain of him if one considered only his humble addresses to other bishops. Nevertheless, there is evidence for a kinder view than was taken by his critics. He was assiduous in visiting the country parishes,[6] and if he assumed that other bishops were saints, he was not blind to the defects of his own clergy.[7] To those of his flock who had sinned he administered a moderate and kindly correction. Few letters of Sidonius give such a happy impression of the man's character as that in which he pleads with a father to forgive the offences of

[1] Claud. Mam., *ap. E.* iv. 2. 3; Ruricius, *Ep.* i. 8. 1.

[2] Greg. Tur., *Hist. Franc.*, ii. 15 (22).

[3] Greg. Tur., ib.

[4] *E.* vii. 9. 14.

[5] *E.* vii. 6. 3, 'non uerebor, etsi *carpat zelum in me fidei* sinister interpres, sub *uanitatis inuidia* causam prodere ueritatis'.

[6] *E.* ix. 16. 2, 'peragratis diocesibus'; cf. *E.* iv. 15. For 'diocesis' in the sense of 'parish' see Hatch in *Dict. Christ. Ant.*, ii, pp. 1554–5. 'Parochia', however, seems to occur in this sense in *E.* vii. 6. 8.

[7] *E.* iv. 9. 5; vi. 2. 2: cf. Dill, pp. 132–4.

a truant son,[1] or that in which he tells us how he remonstrated with a man who had deserted his wife.[2] 'At the sight of the man's repentance', he says, 'I could not confine myself to rebuke, I gave him a few words of consolation.' For a man privileged by his birthright to command, and not by nature kindly disposed to human frailties, those words of consolation cannot have come easily.

To the poor in his diocese he was a bountiful almoner, and we may well believe the tradition that he would often present them with the plate from his table.[3] 'He lives most to his own advantage,' he wrote, 'who does heaven's work on earth by pitying the poverty of the faithful;'[4] and though he uses these words only to praise another bishop, we may in truth apply them to himself. 'I wonder,' writes Claudianus Mamertus, 'if you will ever involve yourself in any interest, which does not turn out to other men's advantage. . . . When you lavish your goods on the poor, there is a sense in which you may be said to serve yourself, but your aim is the service of others.'[5] There may be flattery in these words, but they are not all falsehood. The Gallo-Roman noble and poet, forced by the public will into this new position, made a real effort to conform to the new demands put upon him. 'As best I may', he says, 'I serve God and speak the truth.'[6] The ideal of an aristocrat—a narrow ideal, but not a base one.

His congregation had elected Sidonius as bishop because they needed a protector, and he was soon called upon to fulfil that function. After the defeat of Riotha-

---

[1] *E.* iv. 23.　　[2] *E.* vi. 9.　　[3] Greg. Tur., *Hist. Franc.*, ii. 15 (22).
[4] *E.* vi. 12. 1.　　[5] Claud. Mam., *ap. E.* iv. 2. 3.
[6] 'si aliquid pro uirili portione secundum deum consulas ueritatemque', *E.* vii. 5. 1.

mus, the Breton troops, with all discipline gone, were roaming over the country in disorderly bands. A party of them entered Auvergne and endeavoured to entice away the slaves of a small cultivator.[1] The man appealed to the bishop for assistance, and Sidonius, who had corresponded with Riothamus before,[2] wrote a letter demanding redress.[3] Whether he was successful or no we cannot say, but we can believe that the significance of the incident was not lost upon him. These Bretons were not barbarians, they were the descendants of Roman provincials, allies of the emperor, and they were cantoned among the Burgundians,[4] who remained imperial *foederati*. If such was the conduct of allied troops, what might not be expected from Euric and his Visigoths?

Euric's conduct indeed became ever more menacing, and it may be that his confidence was increased by the knowledge that Ricimer and the emperor were at feud in Italy.[5] There could be no doubt that he was seeking to acquire for himself the greater part of Spain, and in Gaul to extend his territory up to the Loire and Rhone.[6] His Spanish campaigns are exceedingly obscure and the details given to us are few;[7] it seems, however, that already in 468 the greater part of Lusitania was in Visigothic hands,[8] and during the next five years (468–73) most of Spain, with the exception of the Suevic kingdom in the north-west, and the Cantabri and Basques,

---

[1] *E.* iii. 9. 2.   [2] *E.* iii. 9. 1, 'nostri consuetudo sermonis'.

[3] *E.* iii. 9. The identity of Riothamus the recipient of this letter with Riutimus of Jord., *Get.*, xlv. 238, has been generally assumed.

[4] Jord., *Get.*, xlv. 238.   [5] See Cantarelli, p. 97 et seq.

[6] *E.* iii. 1. 5; vii. 1. 1.

[7] For these Spanish wars see Hydatius, 249–50 (ii, p. 35); *Chron. Gall., a. dxi.* 651–3 (i, pp. 664–5); Isidore, ii, p. 281. Cf. Schmidt, i, p. 267; Stein, p. 581.

[8] Hydatius, l.c.

who preserved a savage independence, came into their possession. In Gaul Euric, as we have seen, was taking advantage of the intrigues of Seronatus to detach pieces of *Aquitanica* from the empire, and though Eutropius, who succeeded Magnus Felix as praetorian prefect, as seems probable, in 470[1] was as loyal to the emperor as his predecessor,[2] he could not check the continual advance of the Visigoths. Somewhere about this time the provincials, headed by the Arvernians, resolved to take action against the treasons and extortions of Seronatus, and he was impeached and sent to Rome for trial. Here, however, he was less fortunate than Arvandus, and was put to death.[3] But his absence made little difference to Euric, save that now he had to seize by force what he had previously won by fraud, and perhaps as early as 469, the same year as the battle of Bourg-de-Déols, his troops had overrun those districts of *Narbonensis prima* which were still left to the empire, and occupied Nîmes.[4] Apollinaris and Thaumastus, Sidonius' cousins, were compelled to seek safety in the Burgundian kingdom.[5] In the autumn of 469 Sidonius did, indeed, make a journey into the Rouergue, which seems not yet to have come into Visigothic possession, in order to dedicate a · church.[6] But the outlook was gloomy and the travel-

---

[1] For the date see Appendix D.

[2] *E.* iii. 6. 1. It is very tempting to see in ib. § 3—'Sabiniani familia'—a reference to Seronatus.

[3] *E.* vii. 7. 2. There is no mention of Seronatus being alive after the Clermont incursions have begun, so we may assume that his condemnation is to be dated *c.* 469–70. Fauriel (i, p. 314) comments on the fact that the State scarcely had the courage to carry out the sentence. In explanation of this, he supposes with great probability that Ricimer was an abettor of Seronatus' designs.

[4] See *E.* v. 3. 1, and cf. Tillemont (*Mém. Ecclés.*, xvi, p. 225). This letter, which mentions the election to the episcopate as a recent event, cannot be exactly dated, 469 or 470 are almost equally probable, 471 less likely.

[5] See Appendix C.

[6] *E.* iv. 15. This letter is dated clearly to late autumn ('extremus autumnus

ling dangerous;[1] his host had withdrawn into his forti-
fied mansion in the mountains.[2] He had just built a
new church, and Sidonius writes to congratulate him
upon it; 'a man would scarce dare,' he says, 'at such a
time to repair an old one.'[3] In the next year the
threatened danger had become a reality,[4] and when
Sidonius made his journey to Bourges (470) in order to
conduct the election of a metropolitan, he lamented
that of all the cities in *Aquitanica prima* Clermont alone
was left to the empire.[5] That 'angle' alone stood between the Visigoths and their goal.[6]

When the next year came (471) Euric did not delay
his attack upon it. A force was dispatched, and the
population of Clermont saw again after forty years a
barbarian army before their walls. Each summer for
four years the Goths invaded Auvergne and still the
town maintained an obstinate defence.

We have a considerable number of letters from Sidonius which refer to the events of these years. But to
construct from them an account of the siege which
would satisfy a military historian is not possible. We

—§ 3—cf. 'sub uicinitate brumali'), but the year is uncertain. It is after Sidonius'
election as bishop and prior to the sieges of Clermont. We may choose between
469, when the Goths were not in possession of Rouergue, or 470, when they were
(see below, n. 5). The tone of the letter with its assumption of impending
danger seems to suit 469 better.

[1] 'uiatorum sollicitas aures', *E.* iv. 15. 3.

[2] 'castellum . . . Alpinis rupibus cinctum'.

[3] Ib., § 1.

[4] The only quarter in which Euric suffered a check was in the north-west,
where his troops were defeated by Count Paulus and the Franks near Angers
(Greg. Tur., *Hist. Franc.*, ii. 13 (18)).

[5] *E.* vii. 5. 3. This implies (see *Notitia Galliarum* in Mommsen, *Chron. Min.*,
i, pp. 603–4) that the sees of Bituriges (Bourges), Ruteni (Rouergue), Albigenses
(Albi), Cadurci (Cahors), Lemouices (Limoges), Gabali (Javols, near Mende),
and Vellaui (St. Paulien near Le Puy) were all in Visigothic hands by 470.
Gabali, at least, had been Roman in 469 (*E.* v. 13. 2).

[6] 'angulus', *E.* iii. 1. 4; cf. *E.* iii. 4. 1; vii. 1. 1.

cannot say whether in the various campaigns the attacks were pressed home in different ways; only rarely
can the events described in a letter be assigned even
conjecturally to a certain year. It is a real disappointment, for there is no military action during the latter
half of the fifth century about which we know more,
and yet we know distressingly little. A few general
considerations may, however, be put forward.

The town of Clermont is situated in the broad flat
valley of the Allier known as the Limagne, about eight
miles to the west of the river. It stands upon a large
low hill, which rises about 120 feet from the plain.
But, though actually in the Limagne, the town is only
about a mile to the east of the great chain of Puys
which forms the western boundary of the Allier valley.
This magnificent range of extinct craters twenty miles
long follows the line of the river for an equal distance
to north and south of Clermont. Steep and almost con-
tinuous, there are few points at which it can be scaled
by a road. Three gaps there are, however, opposite
the town, those of Royat, Villars, and of La Barraque.
Roads are carried through all of them at the present
day, and through that of Villars ran in ancient times
the great trunk road from Lyons to Bordeaux.

To the north and south the town is equally over-
looked by lofty hills, for the mountain mass pushes out,
as it were, two bluffs into the Limagne which bound
the view on either side. Of these bluffs that to the
north is formed by the Plateau of Chanturgue, which
is little more than a mile from the centre of the town;
the southern which is more distant is formed by the
hills of Montrognon and Gergovia. Thus the town
stands in the corner of a small oblong plain, bounded

on three sides by heights and making contact to the east with the main valley of the Allier.

From a military point of view the town's position is weak. Artillery, even of a century ago, could batter it to pieces in quite a short time from the neighbouring rises of ground. Nor was its position strong in ancient times either against close or long range operations. The height of the hill on which it stands is too low, and the slope of it is too gentle to present any great difficulty to a storming party. Again, situated as it is at the head of a small plain, it could have been blockaded by quite a small force with very little trouble. An enemy who had only enough troops to hold the narrow mountain gaps,[1] and to intercept foragers and convoys seeking to cross the base of the triangle, between Monferrand and Aubière, a distance of less than three miles of flat ground, could reduce the town to the extremities of starvation.

It was perhaps just because the site was so weak that Augustus chose it to be the cantonal capital instead of the almost impregnable Gergovia. The Augustan town[2] straggled down from the hill on all sides, and its area, as far as we can calculate it from the determination of sites and of the cemeteries which enclose them, can be fixed at not less than 200 acres;[3] the area, however, within which traces of habitation are found does not,

[1] It does not appear that the Royat and La Barraque roads existed in Roman times.

[2] All works on Roman Clermont are superseded by Audollent's careful study in *Mélanges littéraires publiées par la Faculté des Lettres de Clermont-Ferrand* (1910), pp. 103-53; and later discoveries have added little to it. (See *Rev. d'Auvergne* (1912), pp. 70-1; (1921), pp. 144-7; (1927), p. 14.) His work only shows, however, what little evidence there is from which we can determine the history of the town, and how much has been lost through negligence.

[3] This does not include the small thermal establishment of St. Mart, in the parish of Chamalières, about a mile distant.

as far as can be judged from the remains, seem to have been at all completely built over. It does not appear that this large area was ever enclosed with a wall.

Of the Roman walls which are mentioned by Sidonius and Gregory of Tours[1] not a trace remains above ground, and only a few fragments have been brought to light from time to time.[2] From the evidence, such as it is, it appears that there was included within the walls an area of not more than 40 acres.[3]

This disproportion need not surprise us, for it has many parallels in the West. After the great German invasion of 276, which devastated the open towns of Gaul, the inhabitants threw up walls around the small central area in which the administrative offices were situated.[4] The walls enclosed a small area, not only because they were thus easier to defend, but because the importance of the city became more and more identified with that of its executive buildings.[5] Clermont was one of the towns destroyed by the Alamanni in the third century,[6] and thus we may reasonably infer that its *enceinte* belongs to the group of late town-walls.[7]

This does not imply that the parts of the town outside

---

[1] See the passages collected by Audollent, p. 107, n. 2.

[2] Audollent, p. 109 et seq.

[3] In the sixteenth century Clermont was surrounded by a wall which enclosed an area not much less than that of the original town. This wall has also mostly disappeared, but from what remains it seems to be medieval. No antiquary has ever suggested that it was Roman. See Belleforest, *La Topographie universelle*, p. 226 and Plan.

[4] Blanchet, *Les Enceintes romaines*, p. 304.

[5] See Wheeler, in *J.R.S.*, xvi (1926), p. 192.

[6] Greg. Tur., *Hist. Franc.*, i. 29 (32).

[7] Bouillet (*Statistique monumentale du Département du Puy-de-Dôme*, p. 153) alleged that a part of the Roman *enceinte* which he observed was nine feet thick, and that pink mortar was used in its construction. He also reports the discovery of a square tower attached to the wall. All these details are consistent with a late third-century wall.

the new walls had already been deserted, or that they became so at once when the walls were erected. Even though we know that the later centuries of the empire were a period of general depopulation,[1] and particularly of decline in the size of the towns,[2] we must not infer that because a town of 200 acres is found to have erected a wall enclosing only 40 acres, that therefore the population of Clermont had sunk to one-fifth of its former level. We have in fact clear evidence that there were at least monastic edifices outside the walls,[3] and it has been thought that the Mur des Sarrazins, the only relic of Roman Clermont which is still to be seen above ground, dates from the end of the third century.[4]

Such, then, was the aspect of the town. For its immediate defence the inhabitants had to rely mainly upon their own resources; the only allies from whom assistance could be expected were the Burgundians who still remained federates of the empire.[5] A Burgundian garrison did indeed enter the town, but the citizens were not blind to the reasons that had brought it there. The Burgundians, they felt, only wanted to save Auvergne from the Visigoths in order to get it into their own power.[6]

The size of the attacking army is quite unknown to

[1] Seeck, *Untergang*, i, pp. 337–91.

[2] Jullian, *Histoire de la Gaule*, iv, ch. 1.

[3] St. Alyre, Bouillet, *Tablettes d'Auvergne*, p. 604; Greg. Tur., *Hist. Franc.*, ii. 15 (21): St. Saturnin, Chaix, ii, p. 389. Cf. Crégut, *Le Cénobite Abraham*, p. 7, n. 2.

[4] Blanchet, *Enceintes*, p. 164; Audollent, p. 125; *L'Université de Clermont-Ferrand et le Pays d'Auvergne*, p. 106. Jullian (*Rev. Ét. Anc.*, 1913, pp. 423 et seq.) notices that in *Form. Aru.*, i (ed. Zeumer, p. 28), there is a contrast between 'urbs Aruerni' and 'Castrum Claromons'. From this he infers that the area on the top of the hill within the walls was called 'Claromons', and the whole town, including the suburbs outside the walls, 'Aruerni'. He is very probably right. [5] Schmidt, i, p. 380; cf. *E.* v. 6. 2.

[6] *E.* iii, 4. 1; vii. 11. 1.

us: Sidonius speaks in one passage, it is true, of several thousands of them,[1] but he had no means of counting, and it is well known that untrained observers nearly always exaggerate the numbers of an enemy. Both cavalry and archers[2] were included in their force; and, as Sidonius says that even when decapitated a Gothic could be distinguished from a Roman corpse,[3] we may assume that their form of armour was noticeably different.[4] There was at least one Gallo-Roman fighting in the Gothic ranks.[5] If, as seems to be the case, he was a pressed man, we may conjecture that the numbers of Goths available was inadequate, and this will not surprise us when we reflect that Euric was himself campaigning at this time in Spain.

Siege is the title given by Sidonius[6] to the Gothic attacks, at least in 474; yet, as far as we can see, they did not resemble a siege as we should understand the term. It is true that Sidonius mentions the half-ruined and burnt walls,[7] but it is obvious from the letters that

---

[1] *E.* iii. 3. 3, 'aliquot milia Gothorum'.

[2] Gothic cavalry: *E.* iii. 3. 7, 'cunei turmales'. Contingents of horse and foot are similarly mixed in a Visigothic battle-array described by Merobaudes, *Pan.*, i. 20–2. For Gothic foot-soldiers see *Not. Dig. Or.*, v. 61; vi. 61. Archers, Vegetius, i. 20; cf. *E.* i. 2. 5; v. 12. 1. See Schmidt, i, p. 294.

[3] *E.* iii. 3. 7.

[4] The archaeological material for determining the Gothic panoply is collected by Barrière-Flavy (*Les Arts industriels des peuples barbares*, i, p. 49). The finds chiefly consist of the long spears ('Conti') (cf. Greg. Tur., *Hist. Franc.*, ii. 27 (37); Jord., *Get.*, li, 261).

[5] Schmidt (i, p. 263) believes that the Goths were commanded by the Gallo-Roman Victorius. But the passage on which he relies—Greg. Tur., *Hist. Franc.*, ii. 15 (20)—is of very uncertain interpretation. Calminius, *E.* v. 12: his case is exceptional but not unparalleled (see Greg. Tur., *Hist. Franc.*, ii. 27 (37); Stein, p. 568, n. 1, and cf. Schmidt, i, p. 294). There is no reason to suppose (as Fertig, ii, p. 13) that he was actually an Arvernian.

[6] *E.* v. 16. 3, 'Obsidio'; cf. v. 12. 2.

[7] *E.* iii. 2. 1, 'semiustis moenibus'; iii. 3. 3, 'semirutis murorum aggeribus' (it is highly doubtful whether any inference as to the construction of the wall should be drawn from this rhetorical periphrasis); vii. 1. 2, 'ambustam murorum

no attempt was made to storm the town. If the Goths had tried to force the gates or scale the walls and had been repulsed, we should certainly have heard about it in one of the letters. If it is said that the walls were 'scorched' it may be due to the fact that the inhabitants were brought from the lower town inside them, thus leaving buildings exposed to the Gothic firebrands. This seems to be borne out by the fact that the priest Constantius, who visited the town, is said to have wept tears at the sight of burnt houses and ruined dwellings.[1] It is almost inconceivable, as has been stated, that the Goths can have forced their way inside the walls; if these burnt houses therefore had been inside, it would mean that the Goths possessed projectile-throwing engines.[2] It is just possible that they did, for Attila had them at the siege of Aquileia in 452.[3] We do not hear of their use, however, in the course of Merovingian or Visigothic history until a much later date.[4] We may therefore assume that the burning of houses mentioned by Sidonius occurred in the parts of the town outside the walls. The fact that the walls are called ruinous may be explained by supposing either that the Goths had battered them or, more probably, because they were already in a dilapidated condition: it appears that the gaps existing in the wall were filled with wicker-work palisades.[5]

For the most part, however, the Goths seem to have attempted to reduce the city by blockade. When, most

faciem'; vii. 7. 3, 'hoccine meruerunt inopia *flamma*, ferrum pestilentia'; vii. 10. 1, 'semiustas muri fragilis angustias'.

[1] *E.* iii. 2. 1; cf. iii. 3. 8.
[2] Cf. Tillemont, *Hist. des Emp.*, vi, p. 426, n. 1.
[3] Jord., *Get.*, xlii. 221.
[4] Greg. Tur., *Hist. Franc.*, vii. 37; Venant. Fort., iii. 11–12.
[5] *E.* vii. 1. 2, 'putrem sudium cratem' (referring to the siege of 472).

probably in the course of 471,[1] the gallant Ecdicius,
with only eighteen horsemen, succeeded in breaking
through their lines and entering the city,[2] they were
encamped, according to Sidonius' account,[3] in a plain
at some distance from it. This, as has been explained, is the natural action of a foe who wishes to
reduce the city by starvation. Battles between the
Burgundian garrison, aided by those of the citizens who
had arms, and the Visigoths certainly took place,[4] but
from Sidonius' correspondence one is left with the impression that the inhabitants are rather engaged in a
continual watch upon the walls than in fighting,[5] and
this is consistent with the theory of a blockade.

The Goths were not well enough organized to keep
in the field during the winter, and thus when the summer was over they raised the siege and went home.[6]
If a blockade is to succeed under these circumstances,
it is obvious that the crops in the neighbourhood must
be destroyed, or the inhabitants can at once lay in a

---

[1] The date is very uncertain, see Appendix E.

[2] E. iii. 3. 3–5; Greg. Tur., Hist. Franc., ii. 16 (24). Gregory gives him only
ten followers, but he is following Sidonius directly and the discrepancy can
only be due to a slip.

[3] E. iii. 3. 3, 'interiectis aequoribus'; cf. ib. § 4, 'planitie'.

[4] Cf. E. iii. 2. 1, 'campos sepultos ossibus insepultis'; vii. 7. 2, 'cui saepe
populo Gothus non fuit clauso intra moenia formidini, cum uicissim ipse fieret
oppugnatoribus positis intra castra terrori'; vii. 7. 3, 'inopia flamma, *ferrum*
pestilentia'. Most of the populace lacked arms ('turbam inermem', E. iii. 3. 6).

[5] Cf. E. iii. 7. 4, 'peruigili statione . . . muralibus excubiis'; vii. 1. 2 (of
the 473 siege), 'propugnacula uigilum trita pectoribus', cf. E. vii. 7. 2.
One wonders if 'de sorte certaminum si quid prosperum cessit . . . si *quid
contrarium*' (ib.) refers to some defeat of the garrison, which is suppressed in the
correspondence.

[6] This in itself, however, proves that the Goths were in some degree civilized
and settled. Barbarians usually campaigned in the winter. See Seeck, *Untergang*, i, p. 538. We have evidence that the Goths were expected to retire into
winter-quarters in the autumn of 474 (E. iii. 7. 4). The date of E. vi. 6, in which
occurs the phrase 'postquam foedifragam gentem redisse in sedes suas comperi',
cannot be fixed.

stock of grain and neutralize the action of the blocka-
ders. It is natural, therefore, that from more than
one year we should have letters which refer to Gothic
depredations and to the burning of crops.[1] There were
no regular troops in Auvergne to prevent this, and
those of the nobility who were able had fled to the
security of their mountain strongholds.[2] Ecdicius,
however, rose to the occasion; and in 471, at least, his
efforts were successful. Raising a force at his own
expense,[3] he defeated the Goths in several skirmishes
and forced them to retire after remaining barely three
months on Arvernian soil. The gallantry of Ecdicius
actually reached the notice of the emperor, and he was
promised the honour of the patriciate.[4]

For now the Romans at last resolved to protect what
remained of their Gallic dominions;[5] and an army
under Anthemiolus, the son of the emperor, with three
generals all apparently of barbarian extraction was
dispatched to Arles.[6] But Euric, with a Spanish campaign and the siege of Auvergne upon his hands,
showed some disposition to treat. Avitus, a kinsman of
the late emperor, was employed to act for the empire;[7]
perhaps it was thought that a man who bore the same
name as, and was connected with, the old guest-friend
of the Visigothic house would be likely to succeed in his

[1] 'depraedatio', *E.* vi. 10. 1; 'depopulatio', vi. 12. 5; 'segetes incendio
absumptas', ib.; 'populatus', iii. 3. 7.

[2] *E.* v. 14. 1; cf. iii. 3. 7.

[3] Cf. Jord., *Get.,* xlv. 240 ('Arevernam ciuitatem, ubi tunc Romanorum dux
praeerat Ecdicius').

[4] *E.* iii. 3. 7; v. 16. 2. These two passages can be combined without difficulty
to form a narrative as adopted in the text, but the combination is nothing more
than a plausible hypothesis. See Appendix E.

[5] It is probable that the reconciliation between Anthemius and Ricimer
made this expedition possible (Ennod., *Vit. Epiphanii,* 51–75).

[6] *Chron. Gall., a. dxi.,* 649, i, p. 664.   [7] *E.* iii. 1. 4.

mission. It had been for some time rumoured that
Euric would be willing to exchange the newly acquired
territory of *Narbonensis prima*[1] for Auvergne.[2] Sidonius
wrote at this time to Avitus expressing the hope that his
influence would prevent the Romans making any con-
cessions, and then, he said,[3] the Goths, when met
with a firm refusal, would refrain from further demands. What effect this request had upon Avitus we
do not know; but in any case the peace conversations
were unsuccessful, and it is not impossible that Euric
took advantage of them to cross the Rhone.[4] He attacked the Roman army near Arles and completely
defeated it: the generals were killed, and Euric pushed
forward into Provence, ravaging the country. It does
not appear, however, that he remained long in the lands
to the east of the Rhone; perhaps the remains of the
Roman army aided by the Burgundian federates succeeded in ejecting him,[5] and at the beginning of the
next year the Visigothic boundary was again the right
bank of the river.[6] Arles, if indeed it had been captured
at all, became again a Roman town and the seat of the
prefecture.[7]

   The next year, however, brought no hopes for the
imperialists in Gaul; the quarrel between Anthemius
and Ricimer had broken out into open civil war,[8] and,
what was worse, the remains of the Roman army in

---

   [1] Now for the first time in literature called Septimania.   [2] E. iii. 1. 4.
   [3] In the appendix reasons are given for dating the embassy of Avitus to 471.
If E. iii. 1 is combined with *Chron. Gall.* (l.c.), we can then for the first time
understand why Euric was willing to treat at all on such terms.
   [4] *Chron. Gall.*, l.c., '... quibus rex Euricus trans Rhodanum occurrit occisisque
ducibus omnia uastauit'. These words make it certain that the battle took place
on the left bank of the river.
   [5] So Stein, p. 580, n. 3.   [6] See E. vii. 1. 1, and Appendix E.
   [7] On the succession of prefects see Appendix D.
   [8] Cantarelli, p. 102.

South Gaul,[1] which had been brought up to the assistance of Anthemius by Bilimer, the *magister militum Galliarum*,[2] had been cut to pieces before the walls of Rome (July 472).[3] The emperor did not long survive, and in August (472) Ricimer himself in the moment of victory died suddenly of a haemorrhage.[4] In the winter of 471–2 Sidonius looked ahead into the future with an uneasy feeling; earlier in the year he had expressed the wish that he might visit his cousin Apollinaris at Vaison, who had fled before the Goths from his property at Vorocingum and was now residing with his friend Simplicius.[5] But he had a presentiment that something might arise which would make the journey impossible, and indeed the Visigothic incursion had kept him at home.[6] He now heard that Apollinaris had planned for the next year an expedition with the ladies of his family to the tomb of St. Julian at Brioude. Sidonius knew well that Euric would not hold his hand, and that another Gothic incursion might be expected in 472; he therefore counselled Apollinaris not to run the risk.[7] His fears were justified, for the Goths again

---

[1] On Bilimer, 'Rector Galliarum' of *Hist. Misc.*, xv. 4, see below.

[2] 'Rector Galliarum'; so *Hist. Misc.*, xv. 4. Stein (p. 583, n. 1) identifies him with the Vidimer of Jord., *Get.*, lvi. 284 (ΒΙΔΙΜΕΡ-ΒΙΔΙΜΕΡ). This is very tempting, especially as Schmidt's account of the Ostrogoths in this period makes it almost certain that Vidimer arrived in Italy in 471–2 (not as Jordanes, 473–4). But as it is hard to see what meaning the phrase 'rector Galliarum' can have upon Stein's theory, it is probably better to treat Bilimer and Vidimer still as different persons, and adopt Seeck's conjecture (in *P.-W.*, iii, p. 471) that Bilimer was *Magister militum Galliarum*. If it is correct, we must assume that he was appointed in 472 to supersede Gundobad, who had joined Ricimer in opposition to the Emperor (*Chron. Gall., a. dxi.*, 650 (i, p. 664), John Malalas, xiv. in Migne, *Patr. Graec.*, xcvii, p. 557, where ἐκεὶ γὰρ ἦν στρατηλάτης (i.e. *magister militum*) is a slight anachronism).

[3] *Hist. Misc.*, xv. 4.

[4] For these events see Cantarelli, pp. 102–5.

[5] See *supra*, p. 140, and Appendix E.

[6] *E.* iv. 4. 2, and Appendix E.   [7] *E.* iv. 6.

invaded Auvergne. There was more than a mere foray on this occasion, for the invaders advanced to the town and fired some of the buildings, but unable to gain entrance inside the walls, they returned home.[1]

As each recurring year of attack seemed to diminish hope, Sidonius resolved to meet his troubles by prayer and intercession, for he was convinced that the evils which were befalling the city were a punishment for some unknown sin.[2] He had learned that the 'Rogations', a form of prayer instituted at Vienne by Bishop Mamertus, had saved the town from fire and earthquake, and he hoped that they would be equally efficacious for Clermont.[3]

The 'Rogations' were instituted by Sidonius at the beginning of 473;[4] already he knew that another Gothic invasion was expected, already he felt that the resistance of the Arvernians would be unavailing, and that their courage would cost them dear.[5] The Goths came again indeed in 473, and as the year went on he must have felt that his forebodings were justified. In this year Euric himself entered Spain[6] and completed the conquest of *Tarraconensis*, while another army, pushing westwards from *Narbonensis prima*, occupied Arles and Marseilles, and overran Provence.[7] The effect of this new conquest was almost to sever communications between Auvergne and Italy; the defenders could now base no more hopes on Italian aid, though in any case

---

[1] Our knowledge of the siege of 472 is exceedingly scanty. See *E.* vii. 1. 2, and Appendix E.

[2] *E.* iii. 4. 2; vii. 10. 1.

[3] *E.* vii. 1. 2–7; cf. *E.* v. 14. 2.

[4] For the date see Appendix E.

[5] 'animositati tam temerariae periculosaeque', *E.* vii. 1. 2.

[6] *Chron. Gall., a. dxi.,* 651, 652 (i, pp. 664–5); Isodore, ii, p. 231.

[7] See Appendix E and *Chron. Caesaraug.,* ii, p. 222.

the confusion which had followed the deaths of Anthemius, Ricimer, and Olybrius, the successor of Anthemius, would have made any co-operation with Italy almost impossible. Glycerius, who had succeeded Olybrius on the throne, was strong enough, indeed, to beat off a Visigothic invasion of Italy,[1] but he was powerless to take the offensive. Nevertheless, the Visigoths were no more successful than in their attacks on Clermont, and at the end of the summer they were again forced to abandon the siege.

At this point we may pause, and ask ourselves what were the motives which prompted this pertinacious defence. Sidonius writes that it is the 'Rogations' of Mamertus that have put heart into the population, and nerved them to resistance,[2] and though in dealing with a superstitious age we cannot doubt that there was truth in what he said, we may feel that it is an inadequate explanation and one which does too much honour to the bishop to whom the letter was addressed. A circumstance, which at least partially explains their resistance, was the tradition which lay behind them. They boasted that they were descended from the Trojans, and were thus the brothers of the Romans themselves;[3] and men who felt themselves to be partners in the labour of carrying the Latin heritage on their shoulders may well have thought that it was their duty to make no terms with the barbarian. And it may not be unfair to suppose that, as they saw their bishop assisting with heart and soul in the defence, they remembered the words that he had spoken so many years before in the forum of Rome: 'The Arvernians yield to none on foot,

---

[1] *Chron. Gall.*, a. dxi., 653 (i, p. 665).   [2] *E.* vii. 1. 2, 6.
[3] *E.* vii. 7. 2; *C.* vii. 139.

on horseback they can conquer whom you will.'[1] May they not have felt that now the chance had come to prove that he had spoken true? Their bishop, indeed, had another reason for inspiring his flock to resistance. As his thoughts ran over the provinces now subject to Euric they seemed to present a gloomy picture. Euric himself, he had been told, was such a fanatical Arian that he could scarcely keep his countenance when the name of Catholic was mentioned.[2] He was not indeed a persecutor,[3] but he had banished certain bishops, perhaps because he suspected them of treasonable correspondence, and when others had died he had refused to allow the election of a successor.[4] As a result of this, the churches in many places, left without a minister, were falling into ruin. To such a king, Sidonius felt, there could be no surrender, and we can believe that the eloquence of the bishop instilled into his flock a measure of his own sternness of purpose. He himself knew and doubtless he recalled to them the orders of the Gallic Church that no town should be surrendered to the barbarians.[5]

But it is hard for faith and patriotism to contend against starvation. With their crops burnt, and another siege in the next year to be foreseen as the only reward

---

[1] *C.* vii. 149–50.　　　　[2] *E.* vii. 6. 6.

[3] The evidence for a persecution of Catholics contained in Greg. Tur., *Hist. Franc.*, ii. 17 (25); *de Gloria Conf.*, 47, and various hagiographic works is examined and rejected by Yver (pp. 42–6). He points out with justice that even if the circumstances described in *E.* vii. 6 are not exaggerated, they are only temporary. On the other hand Seeck (in *P.-W.*, vi, p. 1239) strangely uses the evidence of *E.* vi. 12, a letter written to Bishop Patiens of Lyons and referring to the *Burgundian* Court, to prove that Euric was less hostile to Catholicism than appears from the correspondence. His article would have gained much from reference to Yver's work (which he does not appear to know).

[4] *E.* vii. 6. 7–9. We must not, however, forget as a corrective to the picture given in this letter, that Euric had permitted the election of a metropolitan at Bourges in 470.　　　　[5] *Conc. Andec.* (453), 4.

for their devotion, we cannot wonder if in the winter of 473–4[1] the hearts of some of the defenders began to waver. There was talk of surrender, and some of the populace were leaving the town.[2] To restore concord Sidonius endeavoured to obtain the aid of Constantius, an eloquent priest of Lyons. Though the winter was hard and Constantius was an old man, he bravely faced the journey from Lyons to Clermont and sought by encouraging language to raise the spirits of the citizens. What arguments he used we do not know, but his speeches were successful and the inhabitants, at unity again, nerved themselves to face a fourth incursion.[3]

All through the summer of 474 the Goths, as usual, lay encamped before the town, and again the crops were burnt. When autumn began, however, the blockade seems to have slackened, and Sidonius, feeling that he could absent himself for a time, set out for Lyons to pay a visit to his youngest daughter, who had been left there with her grandmother and aunts far from the alarms of war.[4] On his arrival he found that the Bishop Patiens had taken upon himself the duty of providing food for the cities of south Gaul which had been devastated by the Goths. He saw a convoy already made up for dispatch to Clermont. Perhaps the Goths had

[1] For the date, which is probable, but not certain, see Appendix (p. 202).

[2] *E.* iii. 2. 2. Kaufmann (*Neues schweizerisches Museum* (1865), p. 13, n. 1) and Yver (p. 32) make much of this passage to show that there was an opposition between the pro-Roman aristocracy and the pro-Gothic plebs. The latter quotes passages from Salvian to show that the Visigoths were welcomed by the lower classes. But this is to throw upon the letter of Sidonius a weight far greater than it should bear. The passages in Salvian describe the results of a definite economic process, and there is no need to assume economic processes in a town that has been besieged for two years. If the *plebs* desired to discontinue the defence we need look no farther than to starvation as a motive. And after all, there was an equal lack of patriotic feeling among some of the aristocracy (*E.* iii. 3. 7).   [3] *E.* iii. 2.   [4] *E.* v. 16. 5.

already retired from the siege by the time that it appeared, for it seems to have entered the town without opposition and brought timely aid for the harassed townsfolk.[1]

Meanwhile the imperial government had begun again to interest itself in Gallic affairs. Glycerius, who had believed that the evils of the empire were due to the simony of bishops, and had tried to abolish it by an edict,[2] now (474?) attempted a more practical measure. There was still a band of Ostrogoths who had entered Italy in 471–2 and, in order to be rid of them, the emperor induced them to settle in Gaul. Quite possibly he had the serious hope that these new barbarians would act as a check against the Visigoths.[3] If he held this hope it was soon shattered, for the Ostrogoths merely joined hands with their Visigothic kinsmen.

Glycerius, however, had little time left him as emperor to institute a western policy. On June 24th, 474, Julius Nepos, the nephew of that Marcellinus who had formerly ruled in Dalmatia, landed at Portus with a strong force provided for him by the new eastern emperor Zeno, and deposed Glycerius.[4]

The new emperor acted with some energy. His quaestor, Licinianus, was at once sent to Gaul with the object of making peace with Euric.[5] He bore with him a letter bestowing on the brave Ecdicius the title of Patrician which had been promised him for his bravery three years before.[6] An army was prepared, which was to march the next spring into Gaul, and if, as has been

---

[1] E. vi. 12. 5–8.

[2] Edictum Glycerii (April 473), in Haenel's Corpus Legum, p. 260.

[3] Jord., Rom., 347; Get., lvi. 284; Sundwall, p. 17; and Stein, p. 583. See above p. 151, n. 2.

[4] Cantarelli, p. 110.   [5] E. iii. 7. 2.   [6] E. v. 16. 1. See p. 149, n. 4.

conjectured, the title conferred upon Ecdicius was accompanied by his appointment to the office of *Magister militum praesentalis*,[1] we may suppose that he was intended to be the commander of it.

Meanwhile Sidonius was still in Burgundian territory. The aggressive preparations of Nepos had led, as it appears, certain Gallo-Roman inhabitants to hope that they might be delivered from the Burgundian rule. An attempt was made to eject the barbarians from Vaison, and to hold it for the new emperor. A group of Gallo-Romans who surrounded Chilperic, the Burgundian king, denounced Sidonius' cousin, Apollinaris, as an accomplice in this affair. One of the last actions of Sidonius in Burgundy before returning was an attempt to establish his innocence before the Burgundian Court.[2]

Towards the end of 474 Sidonius returned to Auvergne: the Visigoths were still in the country, but it was expected that the snows of winter would soon force them to retire.[3] But the miseries of the town had not diminished: when the stores of grain that the generosity of Patiens had provided were consumed, starvation again threatened the inhabitants. Men plucked at the weeds which grew in the crannies of the walls and devoured them. Sometimes in their hunger they ate poisonous herbs and died. A pestilence broke out in the town.[4] We can believe that the thoughts of every man turned to the mission of Licinianus, and all wondered if it would bring peace at last. Sidonius wrote agitated letters to his friend Magnus Felix in the south, begging

---

[1] So Sundwall (p. 18) from Jord., *Get.*, xlv. 241; cf. Ensslin, *Klio*, xxiv (1931), pp. 495–6.

[2] *E.* v. 6; v. 7. 1. See Appendix E.   [3] *E.* iii. 7. 4.

[4] *E.* vii. 7. 3.

for news, and fretted because none came.[1] And when it
came it was not good, for, though we cannot say in
what manner Licinianus' mission was received at the
Visigothic Court, it does not seem to have brought
peace.

In the next spring (475) a small force[2] crossed the
Alps, prepared to dispute with Euric the possession of
Provence.[3] The position was somewhat similar to that
of 471; again Euric expressed a desire for peace, and
again he offered to surrender his latest conquest in
exchange for Auvergne. The latest conquest was now,
however, not *Septimania* but Provence, and the thought
that the Visigoths might at any time invade Italy as
they had done in the previous year may have inspired
Nepos to accept the terms. By doing so, he would at
least ensure his own security. For him the bargain must
have seemed profitable; Auvergne was distant and iso-
lated: to abandon a district over which he held no practical sovereignty in exchange for the wealthy and contiguous Provence was certainly a gain. The discussion
of terms was entrusted to Epiphanius, bishop of Pavia,
and in the late spring he set out for Toulouse. He
seems to have accepted Euric's main proposal, but he
left the drafting of the treaty to four Gallic bishops,
Basilius of Aix, Leontius of Arles, Faustus of Riez, and
Graecus of Marseilles.

It does not appear that Sidonius understood the
nature of the convention which Epiphanius had concluded; for at the beginning of 475 he wrote a letter

---

[1] *E*. iii. 4. 2; iii. 7. 1–3; iv. 5. 2.

[2] Not Nepos' complete army, as is shown from Jord., *Get.*, xlv. 241; *Auct. Haun.*, i, pp. 307, 309.

[3] For the history of the embassies and the events in 475 see Appendices E and F.

offering advice to one of the negotiators, Basilius of Aix. 'These miserable treaties,' he wrote, 'pass through your hands; do your best to obtain for our bishops the right of ordination in whatever limits the treaty prescribes for the Gothic power,[1] so that we may hold them by religion for the Roman church, even if we cannot hold them by treaty for the Roman state.'

From this letter we may realize what Sidonius expected from the peace proposals. The new pretensions of the Gothic king would be allowed, and the old federate relation would now be formally abandoned; in addition to this he expected that Euric's title to his new conquests in Berry, Rouergue, &c., would be recognized. The treaty was a confession of failure, and thus unfortunate, but it was better than nothing. There is no hint in this letter that Sidonius expected his own diocese to be abandoned, and every indication that he did not.[2] Basilius must have felt a twinge of remorse as he read this letter; the sender was indeed to be disillusioned.

And his disillusionment was bitter. When he learnt the terms of the treaty, he saw at once the truth: Auvergne was to be sacrificed for the security of Italy. His people had endured the depths of misery to no purpose; they had been betrayed behind their backs. Sidonius' rage expressed itself in a scornful letter to Graecus of Marseilles.

'Our enslavement', he wrote, 'was made the price for the security of others: the ‚enslavement of Arvernians—the

---

[1] *E.* vii. 6. 10, 'ut populos Galliarum, quos limes Gothicae sortis incluserit, teneamus ex fide, etsi non tenemus ex foedere'. The indicative 'tenemus' shows that 'incluserit' is more probably a future perfect indicative than a perfect subjunctive.

[2] The language of 'populos teneamus ex fide, etsi non tenemus ex foedere', with its use of the first person, is hardly conceivable, from a man who knew that the cession of his own land was an article of the treaty.

shame of it!—men who called themselves brothers of the Romans, men who braved fire, sword, pestilence and famine. Their victories were gain to you, the penalty for their defeats they alone must bear. Was this the honourable peace for which we prayed, when we tore the grass from the walls to sate our hunger? And the end of all this noble devotion is, they tell me, our sacrifice. This peace brings neither honour nor advantage, and I hope that you will regret it. I ask your pardon for the harshness of truth; my distress must excuse my manners. The fact is that you are not interested enough in the general good: when you meet in council, you are concerned less with public dangers than with private fortunes. . . . How much longer is your clever diplomacy to last? Our ancestors will soon be unable to congratulate themselves on any descendants, if they are handed over to a foreign power. You must break this infamous peace at any cost. We are ready to stand new sieges, new battles, new famines. But if we, who were unconquerable by force, are to be abandoned, we shall not forget that this barbarous and cowardly transaction was inspired by you.'[1]

Sidonius was never so great as when he stood out thus as champion of his people and of their Roman inheritance. It was the moment of his life, and he used it well. For in this letter he has written the epitaph of the Western Empire.

[1] *E.* vii. 7. 3–5 (Dalton's translation).

VIII

LAST DAYS

'THE other conquered regions', Sidonius had written, 'have only slavery to expect, but Auvergne must prepare for punishment.'[1] Events proved, however, that this forecast was much exaggerated: indeed, Euric saw that only by conciliation could his dream of a Gallic kingdom be realized: Clermont was not destroyed, and there is no record of its inhabitants being punished. Euric appointed as governor the Gallo-Roman Victorius, who had been Duke in the newly conquered regions of *Aquitanica* and now combined this office with that of Count of Auvergne,[2] residing for the most part, as it seems, at Brioude.[3] Not only did the new government tolerate the Catholic religion, but Victorius actually built churches in Auvergne,[4] and became devotedly attached to the monk Abraham of St. Cyrgues.[5]

Towards the leaders of resistance, however, Euric was not so well disposed. Ecdicius, indeed, had retired to Burgundy before the capitulation,[6] but Sidonius

[1] *E.* vii. 7. 6.

[2] This is a necessary inference drawn by Schmidt (i, p. 292) from *E.* vii. 17. 1 and Greg. Tur., *Hist. Franc.*, ii. 15 (20); *de Gloria Mart.*, 44; *Vit. Patr.*, iii. 1.

[3] The connexion of the castle called 'Victoriacum' at Brioude (Cart. Briu., 252, 339) with Count Victorius, assumed by Boudet (*Les Nationalités en Auvergne*, p. 24), is ingenious and may be accepted as probable.

[4] Greg. Tur., *Hist. Franc.*, ii. 15 (20). Mommsen (*Reden und Aufsätze*, p. 136) strangely supposes that Victorius was an Arian.

[5] *E.* vii. 17. 1; Greg. Tur., *Vit. Patr.*, iii. 1.

[6] Jord., *Get.*, xlv. 240. Ecdicius went to Rome soon afterwards at the request of Julius Nepos. (Mommsen, *ap.* Lütjohann, p. xlviii, alleges, against the evidence, that he was captured by Euric.) It seems from the obscure narrative of Jordanes that he was then superseded in his office of *magister militum* by Orestes (Jord., l. c., cf. Sundwall, p. 18). Nothing more is heard of him, though he seems to have left descendants (Greg. Tur., *Hist. Franc.*, ii. 16. (24), cf. Seeck in *P.-W.*, v, p. 2160). There is no reason for identifying him (as does

remained behind. He was arrested and sent into exile
at the fortress of Liviana (now Capendu), near Carcas-
sonne,[1] and his property was apparently confiscated.[2]

Of the period of his life during which Sidonius was in
exile we have little information. It appears that he was
taken to Liviana on the pretext of some duty,[3] but in
what that duty consisted we cannot say. There is no
doubt, however, that his spirit was broken. His mind
was sick with care, and though at the request of his
friend Leo he undertook a transcription[4] of the life of
Apollonius of Tyana he was unable, he says, to accom-
plish even the most desultory work. The lot of his exile
was hard,[5] his heart was sore,[6] and this inveterate letter-
writer was distressed to learn that messengers were
stopped on the roads and subjected to the annoyance
of a strict examination. 'We must be less assiduous
correspondents,' he wrote to Bishop Faustus, 'we must
learn the art of keeping silence.'[7]

Many years later, Avitus, bishop of Vienne, writing
to Apollinaris, Sidonius' son, reminded him of the
cruel privations that his father had undergone in
captivity;[8] nevertheless, when we search the letters of
Sidonius for details of these privations, we do not find
them.[9] When the imprisoned bishop can only com-

Chaix, ii, p. 188) with Hesychius, the successor of Mamertus in the bishopric
of Vienne.   [1] 'moenia Liuiana', *E.* viii. 3. 1.

[2] *E.* ix. 3. 3, 'patior hic incommoda peregrini, illic damna proscripti'; ib.
§ 4, 'ut . . . non remaneamus terreni quibus terra non remanet'.

[3] See *E.* viii. 3. 1, 'diurna officia'; ix. 3. 3, 'per officii imaginem . . . solo
patrio exactus'. Cf. Hodgkin, ii, p. 308, n. 1.

[4] *E.* viii. 3. 1. On the meaning of 'translatio' see Sirmond (pp. 81–2) and
Mommsen (*ap.* Lütjohann, p. xlix).

[5] *E.* iv. 10. 1, 'me peregrinationis aduersa fregerunt'.

[6] *E.* ix. 3. 3, 'mens nostra saucia'.

[7] *E.* ix. 3. 1–2; cf. ix. 5. 1.

[8] Avit. Vienn., *Ep.* 45.

[9] It is certainly possible that the correspondence has been expurgated, but

plain that his sleep is interrupted at nights by the ceaseless chattering of two Gothic crones, we may believe that to the defender of the Roman language and culture this must have seemed a cruelly ironic insult,[1] but we must feel that, if this is the only horror that he can show us, the tortures which he suffered were rather of the soul than the body.

He remained a captive at Liviana until 476—nearly two years.[2] Deliverance came at last through the efforts of his friend Leo.[3] This man, whose literary gifts had won him the friendship of Sidonius[4] many years before, was now one of the chief ministers of the Visigothic king.[5] Indeed he had drafted for Euric the treaty of 475, and it was appropriate that the man who was indirectly responsible for Sidonius' captivity should be the instrument of his release.

But he was not yet freed from trouble. His mother-in-law, the widow of Avitus,[6] had just died, and left some property to Sidonius. But he found it very difficult to establish his rights. A barbarian had occupied two-thirds of the land, in accordance with the regular

the existence in the collection of such a letter as *E.* vii. 6 makes a theory of expurgation on any large scale not very probable.

[1] *E.* viii. 3. 2. Baret (p. 49) seems to imply that the women were instructed by Euric to annoy Sidonius. 'Le monarque Arien', he writes, 'voulait punir en lui autant le Gallo-romain que l'évêque'. A refinement of cruelty indeed!

[2] So Baret (p. 49) and Mommsen (*Reden und Aufsätze*, p. 136), who points out that the date of the poem in *E.* viii. 9. 5 is fixed by a reference to the usurpation of Odoacer; Fauriel's date of 477 is therefore slightly less probable —though we cannot be certain (i, p. 345; cf. Dalton, ii, p. 179).

[3] *E.* viii. 3. 1, 'cuius incommodi finem post opem Christi tibi debeo'.

[4] *E.* iv. 22. 2; viii. 3. 3; ix. 13. 2, ver. 20; ix. 15. 1, vv. 19–20; *C.* ix. 314; xiv, ep. 2; xxiii. 446–54.

[5] *E.* iv. 22. 3; viii. 3. 3; Ennodius, *Vita Epiphanii*, 85. Cf. Greg. Tur., *de Gloria Mart.*, 91. Gregory calls him 'Consiliarius'.

[6] Mommsen doubts whether the widow of Avitus could have been alive in 476 (*ap.* Lütjohann, p. xlvii, n. 1). I do not see why not.

practice in the Visigothic kingdom,[1] and though Sidonius offered to surrender a further half of what remained to him he was unable to obtain possession.[2]
He resolved therefore to lay his case before the Visigothic king, and with this object he made a journey to
Bordeaux.[3] But the man who had played at dice with
Theodoric found it harder to obtain even the shortest
interview with his successor; for more than two months
he waited in vain. Euric's court was thronged with
petitioners of every nation. During the time that
Sidonius had been a captive the king and his generals
had inflicted defeats upon the Franks[4] and Burgundians,[5] and either in this year or in 477 had again
occupied Provence with Arles and Marseilles.[6] With
Spain almost entirely in his possession,[7] Euric was now
indisputably the most powerful sovereign in the West.
In spite of his distress of mind, Sidonius was attracted
by such a spectacle of greatness. Like many another sub-

---

[1] 'Tertiatio', see Schmidt, i, p. 281.

[2] *E.* viii. 9. 2 ('necdum quicquam de hereditate socruali uel in usum tertiae
sub pretio medietatis obtinui'). The passage is very obscure, but while no
attempts to explain it which do not take into account the 'Tertia Romanorum'
have been at all successful (see Mommsen, *ap.* Lütjohann, l.c.; Dalton, ii, p. 247),
the objections advanced against so referring it are not weighty (see Mommsen,
l.c.; Schmidt, i, p. 281, n. 2), and the comparison which Sidonius makes between
himself and Meliboeus (*E.* viii. 9. 5, ver. 59) supports the view that there was
a Visigoth billeted on his land.

[3] *E.* viii. 9. 1. Tillemont (*Mém. Ecclés.*, xvi, p. 260) holds that Sidonius returned
to Clermont immediately after his release and made the journey to Bordeaux
at a later date. But the fact that he calls himself an exile at the time of his visit
makes this view improbable (*E.* viii. 9. 3).

[4] *E.* viii. 3. 3; viii. 9. 5, vv. 28–30; Schmidt, i, p. 268, n. 3.

[5] *E.* viii. 9. 5, vv. 34–5 (cf. ix. 3. 1; ix. 5. 1); Jord., *Get.*, xlvii. 244; Schmidt,
i, p. 268.

[6] *Cont. Prosp. Haun.*, i, p. 309 (476); *Chron. Gall.*, a. *dxi*, 657 (i, p. 665) (477);
Candidus, fr. 1 (*F.H.G.*, iv, p. 136); Procop., *Bell. Goth.*, i. 12. 20; Jord., *Get.*,
xlvii. 244. The statement of Candidus, in which the deposition of Augustulus
and the revolt of the Gauls against Odoacer seem to be regarded as closely
connected, rather favours 476.   [7] Cf. Jord., *Get.*, xlvii. 244.

ject of the later empire, he had felt a certain reverential awe towards the men who swayed the destinies of their fellows: such a feeling had inspired the official pane-gyrics, and is seen in the letter[1] in which he described his banquet with Majorian. On his election to the bishopric the great metropolitans of Gaul, the spiritual rulers of multitudes, had usurped in his mind the posi-tion which the emperors had filled; but now, as he appeared at the court of the great barbarian king, the old feeling of reverence for worldly splendour returned to him. He had resolved, when he became bishop, to write no more poems: 'Now,' he had said, 'I only think and write of serious things.'[2] But on this occasion he was untrue to his resolution. He enclosed in a letter to his old friend Lampridius a poem of 59 verses in hende-casyllables, the theme of which was the might of Euric.[3] Of all his works this is perhaps the best known. Sido-nius describes how representatives of all the great bar-barian tribes are to be seen as suppliants at the court of Euric. The Roman too is present, and begs that the Garonne, beside whose banks Mars has now come to dwell, may defend the narrow Tiber.[4] Ambassadors even came, he said, from Persia to demand the assis-tance of the Visigoths; and though we may surely here suspect exaggeration, it is fair to point out that the Persians were at this time engaged in a dangerous war at home.[5]

The spectacle of the barbarian nations and even the

[1] *Vide supra*, E. i. 11; cf. Hodgkin, ii, p. 426.

[2] E. viii. 4. 3; cf. ix. 12. 2.

[3] E. viii. 9. 5.

[4] E. viii. 9. 5, vv. 39–44, 'hinc, Romane, tibi petis salutem, / ... / ut Martem ualidus per inquilinum / defendat tenuem Garunna Thybrim'.

[5] An incursion of the White Huns occurred about this date (Schmidt, i, p. 269).

Romans themselves seeking for protection at the court of a barbarian king was indeed a sign that the times had changed. The old idea of an all-embracing empire for which Sidonius had fought was dead even in the breast of its defender. Twenty years before Jupiter had sent the Roman provincial Avitus to save the empire, but now the Italian must realize that Mars has come to dwell in the court of a barbarian king. The parallel of thought between the appeal of the Romans in this poem and the general theme of the panegyric to Avitus is indeed so close that we may suspect that Sidonius had accepted Euric as a representative of Gaul. His feeling of local patriotism, which had always existed alongside of his devotion to the imperial idea, had survived it. The champion of imperial unity had now taken the road that Arvandus and Seronatus had trod: Sidonius had acquiesced in the triumph of separatism. As we read this poem we feel that Sidonius stands between the old world and the new, and is a witness for both of them.[1] That is his unique interest. The unknown poet[2] who related in the epitaph on Sidonius' tomb that 'he had given laws to the fury of the barbarians' was not

---

[1] This aspect of the poem is well brought out by Mommsen (*Reden und Aufsätze*, pp. 136–41).

[2] I do not believe that the Epitaph of Sidonius is to be considered as a contemporary document (cf. Coville, p. 73, n. 1). It seems to imply that *before he became bishop* Sidonius, among other things, gave laws to the barbarians and made peace between warring kingdoms. To neither of these statements do the Letters give any warrant. The Epitaph, as Leblant long ago showed (*Inscriptions chrétiennes de la Gaule*, ii, p. 336), seems to depend upon ps.-Gennadius, 92, and the works of Sidonius himself (cf. 'scripsit uaria et grata opuscula', Gen.—'et post talia dona Gratiarum', *Ep.* 'inter barbarae ferocitatis duritiam', Gen.— 'leges barbarico dedit furori', *Ep.*). It may have been composed in Carolingian times. The Latinity is too good for a much earlier date after ps.-Gennadius, who cannot himself be later than the seventh century (Mommsen, ap. Lütjohann, p. xlvii). For a rather similar instance of a late epitaph perverting the facts, see Epitaph of St. Amabilis in *Bulletin arch. de l'Auvergne*, 1907, p. 191.

indeed accurate in his facts, but his dramatic sense was good.

It may be that Sidonius did not intend this poem merely for the eyes of Lampridius;[1] though Euric himself knew little, if any, Latin,[2] he may well have thought that, if his adulatory poem became known in the Visigothic court, his request would have a greater chance of success. How successful it was, indeed, we do not know: it is certain, however, that Sidonius returned to Clermont soon afterwards as bishop.[3] He soon settled down to his episcopal duties,[4] and was again visiting his friends:[5] moreover, he had the good fortune to gain the favour of the governor Victorius.[6] Nevertheless, he was still in great distress of mind, and the man who had lived for forty years in unchanging faith to the imperial idea must have felt some pain at being forced by circumstances to abandon it. He had lost faith in the world, and trusted only in the hope of a world to come.[7] He wished now, as he said, to be 'delivered from the pains and burdens of present life by a holy death';[8] and his distress was increased by the news that his old friend Lampridius had been murdered by his slaves. His mind was filled with mournful thoughts, and he could neither think, speak, nor write on any other subject.[9]

Though he had congratulated his friend Johannes upon 'deferring the death of literature' by his oratory,[10] he would not aid in the work himself: he was in truth worn out. Kindly friends, realizing that his peace of

---

[1] So Fertig, ii, p. 24.   [2] Ennodius, *Vita Epiphanii*, 90.
[3] Cf. 'reduci', *E.* viii. 3. 2.   [4] *E.* ix. 16. 2.
[5] The letters to Industrius and Vectius (*E.* iv. 9 and 13) are assigned by Chaix (ii, pp. 239, 242) with great probability to this time.
[6] *E.* iv. 10. 2; vii. 17. 1.   [7] *E.* iv. 22. 4.
[8] *E.* ix. 8. 2.   [9] *E.* viii. 11. 14.
[10] *E.* viii. 2. 1.

mind would be restored if only he could find new in-
terests, proposed various literary works that he might
undertake. But their efforts had little success. A request
that he would compose some theological treatise was
met by the reply that he was no theologian, and would,
in any case, have no time to finish it.[1] When Prosper,
bishop of Orléans,[2] asked him to compose the history of
Attila's great invasion and the defence of Orléans by
Bishop Anianus, events which had stirred his youth, he
determined to accede to his request. He actually set
to work on the book, but he soon felt that the task was
beyond his powers and gave it up. The story of Bishop
Anianus and his miracles he did, at least, promise to
complete, but if he succeeded the work is now lost.[3]
Sidonius was no more than forty-five, and yet he was
already considering himself an old man, and abandoning himself to idleness and vague lamentation.[4] His
friends saw the great witness of a vanished age, the man
who had one emperor as father-in-law and who had
supped with another, slipping out of life without a
record of what he had done. His old friend Constantius
begged him to publish to the world a collection of, at
least, his more polished letters.[5]

Now at last he had a real interest to distract his mind.
He fulfilled his promise well, and from 477, at short
intervals of time, seven books of the letters were published. The order and circumstance of their issue are
very obscure. It appears that the first book was published separately.[6] In this a dedicatory epistle to Con-

---

[1] E. ix. 2.   [2] See Duchesne, *Fastes episcopaux*, ii, p. 436.
[3] E. viii. 15.   [4] See E. v. 9. 4.   [5] E. i. 1. 1, 'paulo politiores'.
[6] That Book i was published separately seems certain from the words
'praecipis . . . ut (litteras) omnes . . . *uno uolumine* includam' of E. i. 1. 1. 'Volu-
men' certainly means one book in E. ix. 1. 1; and when Peter (p. 155), who

stantius is followed by ten letters all written before he became bishop.[1] A second book appeared not long afterwards;[2] it contained fourteen letters all similarly prior to his election to the bishopric, but carrying on the story up to the eve of it. These two books would place Sidonius' readers in possession of a document in which the central motive is Sidonius, the great aristocrat, on the one hand a confidant of emperors, kings, and governors, on the other hand a noble living an existence of dignified and cultured ease on his estates. Sidonius' early life, indeed, resembled in its outlines very closely the careers of Pliny and Symmachus, and thus lent itself particularly well to treatment according to the rules of epistolographical form for which they had set the pattern.[3] Many of the letters are genre pictures of Roman life,[4] and, as in the works of his predecessors, some of them contain moral commonplaces.[5] This is not to say that they are not valuable authorities for the history of the times and the character of the man himself; but they contain formalized history, just as one may say that the Georgics contain formalized farming.

believes that Books i, ii, and iii were published together, says that 'uolumen' can mean in Sidonius 'ein ganzes Werk', he can only cite *E.* viii. 16. 1 to prove his point; and it is glaringly inadequate to do so.

[1] Mommsen (*ap.* Lütjohann, p. li) holds that the first book was published at Rome in 469. This is not impossible, but 477, the date adopted by Germain (p. 72), is more plausible, as *E.* i. 1. 4 seems to indicate that a considerable time had elapsed between the publication of the poems and of the letters, and *E.* ix. 16. 3, vv. 49–50, implies that he did not begin to publish the letters until he was bishop. Much less, however, can Mommsen's date of 472 (l. c.) for the second book be sustained. It is very improbable that Sidonius was issuing a book of letters at a time when year after year the Goths were making incursions into Auvergne.

[2] The separate publication of Book ii is to be assumed from *E.* iii. 14. 1, '... quod meas nugas ... confectas opere prosario ... plus uoluminum lectione dignere repositorum, gaudeo ... quod recognoui chartulis occupari nostris otium tuum'. [3] See Peter, p. 151.

[4] Cf. *E.* i. 5; i. 8; ii. 2; ii. 9; ii. 14. [5] Cf. *E.* i. 3; i. 4; i. 6; ii. 13.

They give a picture of a Roman noble, who is *per acci-dens* the man Sidonius, rather than of the man Sidonius who is *per accidens* a Roman noble.

Like most of Sidonius' works, these letters were sharply criticized,[1] but their reception on the whole must have been favourable, for they were soon followed by a third book.[2] In this care is taken to show the reader that he is dealing now no longer with the aristocrat but with the bishop. In the opening letter Sidonius sets out clearly his changed condition. In it he speaks of 'the town of Clermont, whose unworthy bishop I am'.[3] This letter was written, as we have seen, in 471, after Sidonius had already been bishop at least a year. That he actually used such a phrase in the letter which he sent is very improbable. It is a finger-post not for the original recipient but for the reader of the collection. It sets the tone of the book, which commences with nine letters, all subsequent to his election, but prior to his exile. Then follow four apparently from 461–7, inserted perhaps because he could not at the moment find any more letters from the period 469–74 addressed, like the others of the book, to laymen, and the book closes with a formal letter which has the effect of linking it up with the preceding two books.

Soon after the publication of the third book, Sidonius was urged by Leo to compose a history of the times; and perhaps the request was prompted by the desire to have some work which would fit these letters into their proper historical context.[4] But Sidonius' reply was an

---

[1] *E.* iii. 14. 2.

[2] Separate publication is proved from *E.* iv. 10. 2; iv. 22. 1. As both of these letters refer to the exile as a very recent experience they serve to establish the date (*c.* 477) for the publication.

[3] *E.* iii. 1. 2 [4] *E.* iv. 22.

uncompromising refusal; clerics, he said, were not at all fitted for the task of writing history, and Leo himself could undertake the task far better.[1] With his publication of the letters now, as he thought, completed he could return to the ordinary round of life.

He had underestimated, however, the popularity of his work. There was a demand for the production of more letters, and this demand Sidonius met by the publication of Books iv and v.

A fundamental difference may be observed between these two books and those which had preceded. They were published, as it seems, in some haste, for upon examination we find that the verbal reminiscences of Pliny and Symmachus have dropped by nearly a half.[2] The deliberate arrangement of the matter according to a rough chronological scheme, which we have noticed in the earlier books, is now quite absent, and the letters range over a period of sixteen years in what seems to be quite a fortuitous order. Nevertheless, the formal genre pictures in the manner of Pliny, though less common,

[1] *E.* iv. 22. 1, 'ut epistularum curam iam terminatis libris earum conuerteremus ad stilum historiae'. The run of the sentence is against the suggestion that 'iam' should be taken with 'conuerteremus' (Klotz in *P.-W.*, ii. 2, p. 2235).

[2] The details of borrowing from Pliny and Symmachus in each book are appended below: they yield very interesting results.

	Pliny	*Symmachus*	*Average number of reminiscences per ten pages of Teubner text*
i	15	5	7·1
ii	24	6	13·0
iii	8	5	7·2
iv	12	3	4·3
v	6	4	4·6
vi	4	2	5·0
vii	6	1	2·1
viii	11	7	7·4
ix	6	8	6·9

are still to be found in these two books.[1] That they do still appear is not a little remarkable, for Sidonius had become a bishop when many of these letters were originally written. His subservience to the art-form of epistolography forced him to adapt letters which were appropriate to a bishop, and were, according to him, written in a bald, unvarnished style,[2] to suit the character of a dignified aristocrat. The task was not easy, for both of his models were pagans of somewhat conventional devotion; and as a result of this adaptation the old art-form appears in some of the letters slightly moulded to suit the episcopal character of the writer.

It was obvious that letters of such a character could hardly be published as coming from one bishop to another, and thus it cannot be mere chance that the first five books contain no letters to bishops and only two to priests.[3] Accident, however, compelled Sidonius to change his plan. He wished, as it appears, in imitation of Pliny to publish a second triad equal in length to the first.[4] Being unable, perhaps, to lay hands on enough letters to laymen, he resolved to publish a book of sufficient length to fill out the second triad, consisting entirely of letters to bishops. This is the sixth book, which was dedicated to Bishop Lupus of Troyes.[5] In this book the old aristocratic motive had of necessity to be abandoned, and though the theme of Sidonius' humility and

---

[1] *E.* iv. 9, 12, 13, 14, 15, 22; v. 7. 9.

[2] *E.* iv. 3. 9; iv. 10. 2; vii. 2. 1; viii. 16. 3–4; ix. 3. 6; ix. 11. 6. Peter, however (p. 123), believes that these references to everyday style are merely 'die übliche Koketterie der Schule'.

[3] The exceptions are the introductory letter to Constantius, and the correspondence with Claudianus Mamertus. The latter is obviously an imitation of Pliny's correspondence with Tacitus.

[4] I adopt the conjecture of Peter, p. 155, who observes that the length of *E.* i–iii is almost exactly equal in pages to that of iv–vi.

[5] Cf. *E.* ix. 11. 5.

unworthiness is carefully worked out, this short book of twelve letters has on the whole more of an air of spontaneity than its predecessors.

Sidonius had now produced a collection of letters ranging over twenty-four years of his life and embracing a variety of subjects. Nevertheless, his readers were still unsatisfied. Sidonius had not contemplated any further publication, and the supply of letters which he could find in his bookcases was not large. By lengthening such letters as he could find, however, he succeeded in publishing a seventh book. It consists of seventeen letters, of which the first eleven are addressed to bishops; five to laymen follow, and the book ends with a dedication to Constantius. If the number of Pliny and Symmachus reminiscences are a guide, this book would appear to be the least revised of all, and though it contains two very formal moralizing epistles[1] the letters in this book even more than those of Book vi wear an air of freshness: the letters to bishops are more diffuse and varied in matter.

With the publication of the seventh book Sidonius thought that he had really finished at last. He regarded the concluding epistle to Constantius as the epilogue of the whole collection of seven books;[2] 'with you my work began,' he wrote, 'with you it shall end.'[3] Again, however, he was mistaken. His readers asked for more letters and, willing to satisfy them, he searched through his bureaus and published two more books (viii and ix) of sixteen letters each. He had now, as it seems, quite recovered his good spirits; there are no further references to mental distress in contemporary letters. When his correspondents pleaded for a copy of verses from his

---

[1] *E.* vii. 13 and 14.   [2] *E.* ix. 1. 2.   [3] *E.* vii. 18. 1.

hand, he sometimes found an old unpublished poem for them, but sometimes also, though the task was strange to him, he forgot his vow and composed new ones.

These last two books have one feature peculiar to themselves. The publication of the seven books had brought their author fame. He felt that he was assured of posthumous renown,[1] and those who were fortunate enough to be included in the collection as recipients of letters might feel that they too in a humble way could have a share in it. Men who had received no letter from the great man's pen wrote to demand an opportunity of such immortality,[2] and to requests thus seasoned with the flattery that he loved Sidonius was delighted to accede. Now that he was secure of the fate of his letters, he may have felt that they should not go out into the world in such an unpolished state as Books vi and vii. The old motives, the journeys of a messenger, the miserly man, the scenes from country-life re-appear, and there is the same proportion of Pliny and Symmachus reminiscences as in the first triad.

With their formal treatment of stock *motifs*, many of Sidonius' letters seem to be, as it were, out of time altogether, at least out of the time in which their author lived. And, as the most carefully revised books contain the most formal letters, we cannot doubt that this formalism was his chief aim. In realizing it he received powerful assistance from the style that he employed. In the letters the living man is smothered in the conceits of borrowed verbiage. The blunt facts, as set down in Geisler's list of borrowings, illustrate to what extent Sidonius is dressed in the robes of another age.

[1] *E.* viii. 4. 3; 5. 1; 16. 5. [2] *E.* viii. 5. 1; ix. 15. 1; ix. 16. 3.

Besides Pliny and Symmachus, Virgil, Horace, Statius, Lactantius, and Prudentius have been used to form the style of the letters, and it is not only their inspiration but their very words that are used.[1] In Sidonius we cannot say that the style is the man; so often it is many other dead men. Though he is almost the only writer of non-theological works on a large scale who is extant to us from a whole century, and that a century in which language was rapidly changing, neologisms are far from common in his work; his prose, says Baret,[2] is less abnormal both in grammar and vocabulary than that of Tacitus. The education which made a man live in the dreams of the past had indeed a permanent influence upon Sidonius' literary work.

Yet in spite of what seems a desire to avoid the dynamic in his letters, he has given dramatic interest to the period in which he lived, and has drawn the portrait of an age with a line far more firm and sharp than Symmachus before him. His life was passed in times too stirring for him to be able to escape completely from them; and of these times and their manners he was a careful observer. His power of observation is indeed his only literary quality which posterity can still recognize as a merit. As an observer he was singularly detailed and precise: in another age such a man might have made a first-rate naturalist or field archaeologist. It is this quality which makes him so valuable as an historical source. In his description of the Frankish prince, Sigismer, every detail of dress and armour is precisely described;[3] and his portait of Theodoric the

---

[1] See also Schantz in *Philologus*, li, p. 501; Holland, *Studia Sidoniana*, p. 13.

[2] As it is not intended here to discuss the grammar of Sidonius, reference may be made to the exhaustive treatment of Baret, pp. 106–23.

[3] *E.* iv. 20.

Visigothic king[1] makes him live as does no other historical figure of the century. And any one who has tried to give a written description of the dress and appearance of a stranger will know that for an intellect not specially trained to it the precision of observation which Sidonius displays is an achievement of uncommon merit.

Nevertheless, the very quality which makes him so valuable as an historical authority does him harm as a literary man. He saw observation as an end rather than a means: too often he lets his passion for detail run away with him. In an over-scrupulous desire to omit nothing, he allows his readers to forget, and indeed himself forgets, the theme which the details are to illustrate.[2] He never seems to trust either his reader or himself; he says that he will be silent on a point, and then develops it to the full,[3] piling simile upon simile and saying in ten words what he could say in one. Throughout his prose there seems often to appear the rather pathetic figure of a man who lacks the courage to say a straightforward thing in a straightforward way. He is typical of that failure of nerve, that species of intellectual cowardice, which is characteristic of the last productions of the pagan culture.

Sidonius was a true product of a culture which aimed to produce not original thinkers but antiquarian rhetoricians. The antiquary's part in Sidonius' writings we have already noticed, the rhetorician's is equally obvious. The culture of the age in which he lived demanded from its literary men a perpetual striving

---

[1] *E.* i. 2. For other specimens see *E.* iv. 9 and 13, and cf. Dalton, i, p. cxxvii.

[2] Conspicuous examples are *E.* iii. 13; v. 7; vii. 14; viii. 14.

[3] *E.* iii. 3. 1–2; iii. 13. 7–9.

after stylistic effects. A great writer might have reacted to this demand in two ways. He might have defied convention altogether. In that case his works would probably not have survived to us at all, a point that detractors of Sidonius are apt to miss. He might, on the other hand, have let his imagination play around the conventions, and illuminated the conceits that they demanded with flashes from his own intelligence. Sidonius did neither of these things; he lacked the depth of mind to give to his conceits an intellectual appeal. His metaphors may be far-fetched and strange, but they do not illuminate;[1] the antitheses of which he was so fond are too often the formal antitheses of words rather than of the thoughts represented by words.[2] In his search after formal antitheses he often, indeed, took the line of least resistance, the unashamed verbal pun: Perpetuus' church must last Perpetually,[3] and the bishop who failed to receive Sidonius' letter at the town of Apt is said to have Aptly escaped its perusal.[4] When Sidonius desired to rise to a height of eloquence above his usual vein his method was to express his meaning in words of more than usual length.[5] The view that long words are an indication of noble writing is commonly a phase of youthfulness, but in Sidonius it was the symptom of a universal malady, the second childishness of a civilization. With pardonable petulance his sympathetic interpreter Dalton exclaims that such language as his calls aloud for the

[1] Cf. *E.* i. 7. 12; iv. 13. 4; ix. 11. 9; Dalton, i, p. cxxxi.

[2] Peter (p. 153) well quotes *E.* viii. 7. 2, 'quique sic uitiis, ut diuitiis incubantes', and ix. 4. 2, 'scientia fortis, fortior conscientia'. See also Dalton, i, pp. cxxviii–cxxix, and Fertig, iii, p. 17.

[3] *E.* iv. 18. 5, vers. 20.

[4] *E.* ix. 9. 1.

[5] Cf. *E.* i. 1. 4; i. 5. 11; ii. 2. 5, &c.

knife.[1] But it cannot have it, for the man is the product of his age and cannot be separated from its conventions.

Sidonius' ninth book was published, as it seems, *c.* 479–80,[2] and after it we have no word from his pen. He had made the promise, indeed, that he would write more verse in honour of Christian martyrs, but if he kept his word the poems have not been preserved. For the last days of his life we depend upon the traditions preserved by Gregory of Tours. They may contain truth, but we cannot control them.

Quite soon after the publication of the letters, Sidonius was smitten with a grievous blow. Victorius, the Count of Clermont, whose tenure of the office had begun so auspiciously, began to behave in a tyrannical manner. He became licentious and cruel, and endeavoured to assassinate Eucherius,[3] an Arvernian noble, who had been a correspondent of Sidonius in earlier days. The populace rose in anger against him, and he fled to Rome. With him went Apollinaris, Sidonius' only son.[4] The lad had shown himself backward in his studies,[5] but Sidonius had loved him well, and must have been sadly distressed when he absconded with such a disreputable man. Victorius' conduct in Rome was so scandalous that he was put to death, and young Apollinaris was imprisoned at Milan, 479 (?).[6] He

---

[1] i, p. cxxviii.

[2] The evidence for date is contained in *E.* ix. 12. 2—where Sidonius, in answer to a request for a poem, uses the words 'postquam in silentio decurri tres Olympiades.' As this must refer to the publication of the poems in 469, the letter is dated to 480. This may, however, be a round number, as the 'annos circiter uiginti' of *E.* ix. 13. 6 (459–79) certainly is. As we shall see, there is some reason for conjecturing that Victorius took the young Apollinaris with him to Rome in 479, and if so *E.* ix. 1 and with it the whole book must be prior to that event.   [3] Greg. Tur., *Hist. Franc.*, ii. 15 (20).

[4] Greg. Tur., *de Glor. Mart.*, 44.   [5] *E.* ix. 1. 5.

[6] The date is uncertain. According to Gregory (*Hist. Franc.*, ii. 15 (21)),

escaped eventually, thanks to a miracle, according to Gregory's story, and returned to Auvergne, but whether he saw his father again alive we do not know.

It appears from the account of Gregory[1] that Sidonius must have fallen into dotage after this, for two priests[2] succeeded in depriving him of all authority over the church property. Only divine intervention, says Gregory, which caused one of them to die suddenly, prevented them from dragging him out of his own cathedral.

Some time after this Sidonius was taken ill of a fever. Realizing that his end was near he requested that he might be carried into his cathedral church, and there amid the tears of his congregation[3] he died (Aug. 21st, between 480–90).[4] The last words from his lips were a request that Aprunculus, bishop of Langres, might be appointed to succeed him. Though he had never professed any power of working miracles, he received shortly after his death the honour of canonization.[5]

Victorius was made Duke of the Seven Cities in the fourteenth year of Euric's reign. He ruled nine years in Auvergne, and then retired to Rome, where he was executed four years before the death of Euric, who, according to Gregory, reigned twenty-seven years. The dates are consistent with each other, but not with the facts, for Euric reigned seventeen or eighteen years (Schmidt, i, pp. 259, 270, n. 2). The discrepancy can be explained in more than one way, but it seems plausible to suppose that Gregory has endeavoured to harmonize the details of Victorius' life with a source that gave Euric a reign of 27 (xxvii) instead of 17 (xvii) years (466–83). Thus Victorius will have been appointed governor in 471—a very plausible date in itself—and will have gone to Rome in 479.   [1] *Hist. Franc.*, ii. 15. (23).

[2] Chaix finds names—Honorius and Hermanchius—for these priests (ii, p. 366), but he quotes no authority, except Gregory, who gives no names, and I have not been able to discover his source.

[3] Greg. Tur., *Hist. Franc.*, ii. 15 (23). Dalton, however (*Gregory of Tours*, ii, p. 496), points out that the scene of the weeping congregation is a stock motive in Gregory.

[4] For the date see Appendix C.

[5] Chaix, ii, p. 374.

His bones were laid in the church of St. Saturninus, a short distance to the south of the town. When in the tenth century this church was demolished they were transferred to the Basilica of St. Genesius in the western suburbs,[1] where Tillemont saw them in the seventeenth century, preserved, as he says, with piety and little magnificence. In 1794 the church was destroyed, and such bones as could be found in it were burnt in the Place de Jaude.[2] But there was an old tradition that he was buried not at Clermont but in the church of Aydat,[3] and there, it may be, he rests 'quiet', as his epitaph tells us, 'amid the storms of the world'.

[1] Ib. ii, p. 389; Tillemont, *Mém. Ecclés.*, xvi, p. 274.
[2] Baret, p. 102.   [3] Coville, p. 74.

WAS SIDONIUS IMPLICATED IN THE
CONIURATIO MARCELLIANA?

UNTIL the appearance of Coville's careful study (*Histoire de Lyon*, pp. 56–8), writers had not doubted that Sidonius was implicated in the 'Coniuratio Marcelliana'. Coville's theories, however, compel re-examination of the evidence. The question is not as easy as had been supposed; and, even if we do not agree with Coville's conclusions, we can be grateful to him for directing us towards difficulties that had been unnoticed before. So complicated indeed is the problem that, in fairness, the evidence must be placed before the reader for his own examination.

(A) In the preface (*C.* iv) to the panegyric of Majorian Sidonius compares himself to Tityrus who was granted life and lands by Augustus ('praestitit afflicto ius uitae Caesar et agri', v. 3), and to Horace whose life was spared after the battle of Philippi. He pleads for similar treatment for himself from Majorian. Now follow the crucial lines (vv. 11–14):

> sic mihi diuerso nuper sub Marte cadenti
>   iussisti erecto, uictor, ut essem animo.
> seruiat ergo tibi seruati lingua poetae
>   atque meae uitae laus tua sit pretium.

There is no actual internal evidence to date this poem, but it may be fairly presumed to be contemporary with the panegyric, and can be assigned therefore to the end of 458.

(B) Towards the end of the panegyric of Majorian (*C.* v. 574–96) Sidonius pleads for indulgence towards the town of Lyons which had received a Burgundian garrison: notable are the words:

> fuimus uestri quia causa triumphi,
> ipsa ruina placet. (vv. 585–6.)

It may be noticed that Sidonius, in using the plural of the first person, identifies his fortunes with those of the citizens of Lyons (cf. vv. 575, 584). He goes on to say: 'When you celebrate your triumph at Rome, I will sing that you have conquered Africa—ante tamen uicisse mihi.' Then follows another passage of crucial importance (vv. 596–9):

> quod lumina flectis
> quodque serenato miseros iam respicis ore,
> exultare libet: memini, cum parcere uelles,
> hic tibi uultus erat; mitis dat signa uenustas, &c.

The poem can be dated, as has been said, to the end of A.D. 458 (see p. 46, *supra*).

(C) In an undated poem addressed to Marjorian (*C.* xiii), immediately after a request for a reduction of an extraordinary tribute, occur the words:

> Has supplex famulus preces dicauit
> responsum opperiens pium ac salubre.
> ut reddas patriam simulque uitam
> Lugdunum exonerans suis ruinis,
> hoc te Sidonius tuus precatur. (vv. 21–5.)

In the letter written between 461 and 467 to Montius (*E.* i. 11) there is more than one important reference:

(D) § 3. 'Catullinus . . . cum semper mihi tum praecipue commilitio recenti familiaris . . .

(E) § 6. 'Cumque de capessendo diademate coniuratio Marcelliana coqueretur, [Paeonius] nobilium iuuentuti signiferum sese in factione praebuerat.'

(F) § 7. '[Paeonius] multorum . . . odia commouit . . . mihi, adhuc amico.'

Looking through the first three of these passages, we see that in (C) Sidonius makes a request for life and lands, speaking as an inhabitant of Lyons. In (B) he makes a similar appeal for relief in the name of the whole community of Lyons, prefacing the appeal by the statement in (A) that he had himself been pardoned personally. From these passages it has been inferred that after the battle of Placentia Sidonius returned to Auvergne, that he took part in the 'coniuratio Marcelliana' and was punished for his part like the other inhabitants of Lyons by a heavy fine, and that he obtained relief for himself and the men of Lyons by poems delivered to Majorian—*C.* iv. v. and xiii, in fact. This we may call the 'orthodox view';[1] and so well established was it that, though there were differences on points of detail, it was never seriously controverted and consequently the arguments for it were never very carefully set out.

It was observed, however, by Coville that (B), vv. 597–8, show that Sidonius had already been pardoned some time before the panegyric had been delivered. His inference, which is quite new, is perfectly sound. But from it he constructs the theory that Sidonius' pardon mentioned in (A) is to be referred not to the 'coniuratio' at all, but to the battle of Placentia in 456, in which he supposes Sidonius to have taken part. Sidonius, in fact, was pardoned according to Coville's view for an act quite unconnected with the 'coniuratio' and consequently there is no proof that he took any part in it. Furthermore, Coville cites the passage (D) as definite evidence to prove that Sidonius had taken part in a battle. But this, as will readily be seen, is not justified.

Nevertheless, Coville's theory would be entitled to hold the field in spite of its lack of support, if the traditional view could be shown to be impossible. But it should already have been apparent that, so far from overthrowing it, the passage adduced by Coville actually supports it, at least in the modified form in which I would advance it. The theme of (C) is: 'Pardon me!' that of (A) is: 'I have been pardoned'; that of (B) is: 'Pardon in like manner the inhabitants of Lyons!' If this explanation of the three passages be accepted, we may say that we knew already that Sidonius had been pardoned by Majorian before the panegyric was delivered; and thus the statement of *C.* v. 598–9 merely tells us again in different words what we knew already from passage (A)—•

---

[1] See for example, Dalton, i, pp. xx–xxiii.

namely that the request conveyed in (C), so far as it concerned Sidonius himself, was successful. The only objection
which may be taken to this view is that *C*. xiii is sometimes
assigned to a later date than the panegyric, and Coville himself so assigns it. But the arguments for doing so are very
weak. Coville says that there is inadequate time between
Majorian's arrival at Lyons and the date of the panegyric for
the writing of this poem: this argument entirely begs the
question; I maintain that, *from the argument derived from C. v.
597–9 which Coville himself has supplied*, we should hold that
C. xiii was written when Majorian was on his way from Italy
to Lyons. Coville further suggests that Sidonius would not
have called himself 'Sidonius tuus' to the emperor until after
the panegyric. This is not an argument which can be matched
against the fact that Sidonius asks for 'vita' and 'patria' in
(C), and when in (A) he thanks the emperor for having saved
him, he compares himself to Tityrus, to whom 'vita' and
'agri' were given. The similarity between the two passages
is exact; and when Coville, in speaking of *C*. xiii, talks of the
'absence de remerciements du poète', one can only wonder
that he did not see where they were to be found.

With the modification that Sidonius was pardoned before
Majorian came to Lyons we may accept the traditional view
as against Coville's theory, because it provides an explanation of greater simplicity and does not involve an additional
hypothesis. This implies, not that Coville is wrong in saying
that Sidonius fought at Placentia, but that we can justify the
traditional view by showing that the evidence is consistent
with it. Coville's hypothesis cannot, indeed, be refuted, because there is no evidence; but for the same reason it cannot
be sustained.

It may be supposed that, if this proof is cogent, we have
already solved the question. In a sense we have, but in a
sense not. The 'coniuratio Marcelliana' involved two
measures: (i) the attempt to raise to the throne Marcellus or

Marcellinus (whoever he was); (ii) the billeting of Burgundians in and around Lyons. For the latter, every landowner in the Lyonnais might be considered responsible, whether he had actually favoured the measure or not: from this point of view we can hardly doubt that Sidonius was implicated. But whether he took any part in the attempt at setting up an emperor is a question that cannot be certainly answered. We see, however, from (E) that the conspiracy was instituted by the young nobles, and Sidonius was at this time a young noble himself: it was managed by Paeonius, and Paeonius, as (F) tells us, was a friend of Sidonius, in spite of his low birth and unattractive character.

As I have tried to show in the text, the 'coniuratio Marcelliana' was primarily directed against Ricimer; and thus both on *a priori* grounds and from the scanty evidence that we possess, we may consider it not as certain, indeed, but as highly probable that Sidonius took a real part in it in the fullest sense. But the fact that Sidonius hurried to meet Majorian before he reached Lyons, and that he asked for grace and received it, rather suggests that, like other nobles, he abandoned active participation in the conspiracy on the accession of Majorian. But the Burgundian settlement he could not annul, for the barbarians had forced his hand.

APPENDIX B

AVITACUM

A SHORT poem (*C.* xviii) from Sidonius' pen has told us how the name of his villa—Ăvītācum—was pronounced, and in the Epistle to Domitius (*E.* ii. 2) he has given us topographical details which would appear sufficient to indicate the site of it. Yet the investigator who approaches the problem with open mind soon finds that it is much more complicated than he imagined. It is not perhaps of very great importance, and would only interest the general historian if the site could be definitely located in a place where excavation could be done; and this is just what it seems impossible to do.

Nevertheless, it may be of use to have the difficulties clearly stated, if only to warn a future investigator against too willing a reliance upon the enthusiasm of local topographers. For, on the whole, they have approached the subject by a wrong road. We know that Sidonius' villa was situated by a lake, it was an heirloom from the family of the Aviti,[1] and the Aviti were Arvernian nobles. Consequently it is a very probable, though not an absolutely certain, inference that it is to be located upon the shores of a lake in Auvergne.[2] For the most part investigators have decided on a lake and then tried to force the details given by Sidonius into harmony with the landscape. What we should rather do is first to construct an ideal map of the villa and then see whether the topography will agree with it.

The lake, says Sidonius, is seventeen stadia long by nautical measure. We cannot doubt that Sidonius is referring to the normal Alexandrian stadium of which eight were equal to the Roman mile. Our lake will then be slightly over two miles in length. It is fed by a stream which reaches the lake after flowing over a cascade, and brings down such a quantity of alluvium that at its entrance a quagmire has been formed, where it is impossible to walk on foot. The lake extends in a devious course towards the east and at its exit the stream flows underground. The lake is not the mere widening of a stream; it is large enough to be called 'Pelagus' and 'Aequor;[3] and (if we may, for the moment, assume Sidonius' direction-

---

[1] *E.* ii. 2. 3.

[2] *C.* vii. 148–57. Dumas-Damon (*Rev. d'Auvergne*, 1895, p. 301) argues that because Domitius, the first recipient of the dedicatory poem *C.* xxiv lived at Brioude (*C.* xxiv. 16), therefore Avitacum was near Brioude. This argument has no weight, for we do not know whether *C.* xxiv was dispatched from Avitacum.

[3] *E.* ii. 2. 12, 16, 18; *C.* xviii. 8.

LAC D'AYDAT

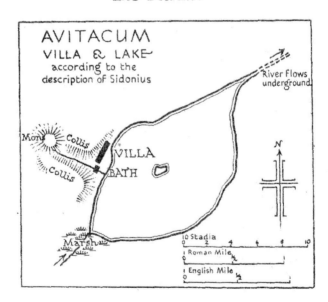

AVITACUM
VILLA & LAKE
according to the
description of Sidonius

points to be exact) it has a north, a south, and a south-west side. Finally the lake has an island in the middle, around which it was possible to race boats.

Moreover, there are data from which we can place the villa itself in relation to the lake. The villa was not actually on the shore, but was near enough to it to receive the spray which a storm might throw up. Its portico looked out upon the lake which was to the east of it, that is to say the villa was situated near its western shore; and it had two wings, like many 'Celtic' villas. It was situated at the mouth of a valley, which descended from the flank of a hill (mons terrenus'), and widened out as it approached the lake, so that a little plain was formed of sufficient size to contain the villa and the bath-building to the south of it. The normal width of this valley, before it broadens out, is 480 feet,[1] and it contains a stream capable of supplying the bath-building, whose cold bath could contain about 20,000 modii of water.

Working upon these data we may construct the ideal map somewhat as that facing this page.

The identification of the lake has interested scholars ever since Sidonius' work became the object of detailed study at the Renaissance; and in the sixteenth century Joseph Scaliger suggested the lake of Geneva. This was soon seen to be impossible, and in 1599 the famous Clermont antiquary and editor of Sidonius, J. Savaron, proposed a site on the west bank of the Lac de Sarlièves,[2] a mere situated in the Limagne under the slopes of the Plateau de Gergovie. It was drained in the seventeenth century, but its outlines can still be discerned. This is more plausible, for whereas the lake of ·

---

[1] The language of Sidonius ['colles quattuor a se circiter iugerum latitudine abductos' (v. l., obductos)] makes it clear that he is using the 'iugerum' as a measure of length, in the same way as the 'arepennis' was used (Greg. Tur., *Hist. Franc.*, i. 6). Plin., *N.H.*, iv. 2. 31 is a close parallel, and shows that 'iugerum' as a measure of length refers to the short side of the oblong 'iugerum' surface (i.e. 120 ft.). See Hultsch, *Griechische und römische Metrologie*, p. 62, n. 13, where this passage from Sidonius is not noticed.

[2] In his edition of Sidonius of 1599, p. 103.

Geneva is forty miles long, that of Sarlièves was about $1\frac{1}{2}$.[1] Nevertheless, it is certain that Sarlièves was fed by a stream from the south, which left it at the north end. There never could have been rocks or a cascade at the entrance. Belleforest's sixteenth-century plan of the lake shows no island. Moreover, on Savaron's theory, the 'mons terrenus' would have been the hill of Gergovia itself, a fact, as Crégut points out,[2] that one would not have expected Sidonius to omit. Difficulties also arise when we attempt the valley 480 feet wide, but to examine them in detail would be to slay the slain.[3]

Savaron's great follower Sirmond[4] pointed out that Sarlièves was inadmissible and proposed somewhat tentatively a site on the Lac de Chambon. The theory was eagerly taken up by M. Bertrand in the early nineteenth century,[5] and commended itself to the naturalist H. Lecoq, who, if he did not know Sidonius well, at least had an unequalled knowledge of Auvergne. The theory, then, is entitled to be heard with respect. But it raises great difficulties. Chambon is not more than a mile long, and it is hard to believe that in historic times it was ever larger. Moreover, when an attempt is made to discover the valley 480 feet broad and the 'mons terrenus' they cannot be found; and the defenders have implicitly admitted this in assuming that the villa is buried under the medieval castle of Murols. But Murols stands on a hill 150 feet above the lake.

With the exception of the Lac d'Aydat, to which we shall presently refer, no other Arvernian lake has found any ac-

---

[1] A. Vergniette has tried to show recently from geological arguments that in the time of Caesar the Lac de Sarlièves did not exist (see *Rev. d'Auvergne*, 1931, p. 266). If he has not proved his case, he has at least shown that it was at one time much smaller.

[2] 'Avitacum', in *Mém. de l'Ac. de Clermont-Ferrand*, 1890, p. 38.

[3] Savaron also used the argument that Aubière was originally Aubiac and Abitac. This strange nonsense has actually gained admittance into Dalton's note (ii, p. 222).

[4] Ed. Sidonii, p. 26.

[5] *Recherches sur les eaux du Mont-d'Or* (1823), p. 499.

ceptance. This is not surprising, for such lakes as Tazenat and Pavin are eliminated by their smallness alone. Nevertheless, it must not be forgotten that numerous 'anciens étangs' have existed in Auvergne; some of them, like the Lac de Sarlièves, have been drained completely, others again are of recent creation, formed by the erection of a 'barrage' across a stream in order to reclaim marshy soil. The exact determination of the original size of these 'étangs' has never been made, and to have attempted it would have entailed on my own part a long and detailed study which is not at all justified by the importance of the result. It is perhaps worth mentioning that the Étang de Tyx (parish of La Celle), which is the largest existing sheet of water in Auvergne, is about $1\frac{1}{2}$ miles long and has an island in the middle. There is a valley leading down to its western bank, which fits the topographical details well enough, and close to it is the church of St. Avit, which (with the exception of a parish church in the town of Issoire) seems to be the only church in Auvergne with such a dedication. It is fair to say, however, that when I visited the site no traces of ancient habitation could be seen.[1]

And finally there is the Lac d'Aydat. Its identity with Avitacum seems first to have been suggested in the eighteenth century by Audigier, Canon of Clermont-Ferrand.[2] His arguments, indeed, were quite indecisive, but the suggestion has found much favour with local antiquaries, and was vigorously supported in two articles by the Abbé Régis Crégut.[3] In these articles there is much special pleading, and points which are unfavourable to Crégut's thesis are apt to be concealed in verbiage. Nevertheless, they have

---

[1] Dr. Canque of Clermont-Ferrand has suggested to me in conversation the Étang du Roi. It is just over the eastern border of Auvergne, but that, as has been said, is not a fatal objection. I have not visited the site.

[2] His 'Histoire de Clermont' in MS. is in the town-library; I owe the knowledge of his arguments to Crégut, 'Avitacum', p. 57.

[3] 'Avitacum' in *Mém. de l'Ac. de Clermont-Ferrand* (1890); 'Nouveaux Éclaircissements sur Avitacum', in *Bulletin de l'Auvergne* (1902).

carried fairly general conviction both in Auvergne and outside.

In one respect the Lac d'Aydat differs from all other sites: its claims can be supported by external evidence. Cartularies quoted by Crégut[1] bring the name Aidacum back as far as the tenth century, and the equation of Avitacum and Aidacum is accepted definitely by Longnon,[2] and after hesitation by Skok,[3] who points out that the disappearance of *v* before accented *i* is abnormal (cf. Avessac, St. Vit). Furthermore, in the Romanesque church of Aydat is a small reliquary on which is the inscription: 'HIC S̄T DVO INNOCENTES ET S̄ SIDONIVS.' The inscription is dated by Dom Morin[4] to the twelfth century, and he has given reasons for supposing that it has been copied from a Merovingian inscription of the eighth century. If he is right the connexion of Sidonius with Aydat can be brought back to a date two hundred years earlier than the cartulary, perhaps only two and a half centuries after Sidonius' death. This is imposing evidence, and the internal difficulties of the identification with the Lac d'Aydat would have to be very great if it were set aside. Yet there are difficulties, and they are twofold. The data of Sidonius are not entirely suitable to the Lac d'Aydat, and even if they were it is difficult to state where on this theory the villa of Sidonius is to be placed in reference to the lake.

We notice at once that Lac d'Aydat is not seventeen stadia in length. The longest straight line joining two points on it is, according to the Carte d'État-Major, slightly under a mile and a half. Nevertheless, it is probable enough that some of the alluvial deposit brought down by the stream by which the lake is fed may have been formed since Sidonius'

---

[1] 'Avitacum', p. 75.   [2] *Les Noms de Lieu de l'ancienne France*, p. 82.

[3] *Die in -ascum, -auscum, -oscum, und -acum, gebildeten südfranzösischen Ortsnamen* (Halle, 1906), p. 63.

[4] *Mém. de la Soc. nat. des Ant. de la France*, lvi (1897), p. 45. I owe this reference to the kindness of M. P. F. Fournier.

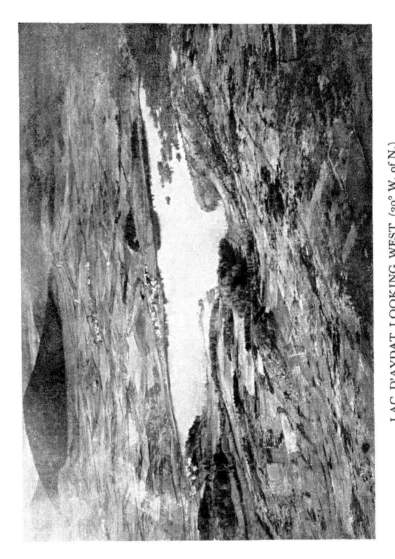

LAC D'AYDAT LOOKING WEST (30° W. of N.)

By permission of the Aéro-Club d'Auvergne

time, and it is a point in favour of the Lac d'Aydat that allusion is made by Sidonius himself to such deposits. We might say that the length of the lake was at least a furlong greater fifteen hundred years ago, and explain the remaining discrepancy as one of those over-estimations of distance which are not uncommon in untrained observers.[1]

Sidonius tells us that there was an island in the middle of the lake. At Aydat there is a group of rocks close to the north shore. The largest of them is called the Île de St. Sidoine, but the name is of recent origin. The air-photograph shows that this island is very far from being in the middle of the lake; in fact it is possible for a man standing on one of the promontories of the indented north shore to throw a stone over to it. To race a boat round it would not be easy, indeed the proprietor of the Hôtel du Lac who has experience of boating on the lake told me that he believed it impossible. The island would only be near the middle of the lake if its waters had extended over the lava-stream or 'cheyre' to the north; but then it must have left some alluvial deposit, and nothing of the kind is to be found. The position of the island is very hard to reconcile with the statement of Sidonius: it is certainly situated on a line bisecting the Lac d'Aydat from north to south, but it is not in the middle of the lake. If it is felt that the strength of the external evidence makes it necessary to turn this objection, I can only suggest that Sidonius, who, after all, was writing an epistle and not a regional survey, may have permitted himself one of those inaccuracies which ordinary men often commit in conversation for the sake of tidiness and succinctness.

The other topographical details do suit Aydat quite well. The lake does extend in a devious course to the east; the stream as it enters does tumble over a cascade and bring

---

[1] Crégut, 'Avitacum', pp. 16, 82, remarks that the ancients always coasted when they made sea-voyages (which incidentally is not true): he supposes therefore that when Sidonius says 'lacus procedit in xvii stadia' he refers to the semiperimeter of the lake.

down a quantity of alluvium; after leaving the lake it does lose much of its waters in the porous lava, and this is not a common phenomenon.[1]

Yet difficulties no less serious confront us when, assuming that the lake is really Aydat, we attempt to locate Avitacum upon it.

There are three valleys which lead down from the west to the shores of the Lac d'Aydat, and at the base of one of the three Sidonius' villa must have been situated: no other site is possible. The most northerly is that of Sauteyras, which was chosen by Dumas-Damon;[2] the central one has never been suggested; the southern, through which runs the main stream which feeds the lake, is that adopted by Crégut. Both Dumas-Damon and Crégut have written at length on their respective sites. Each glosses over the data that tell against his theory, and it is fair to point out that neither has thought fit to mark on his map the exact spot where the villa is to be placed.

1. The Sauteyras site corresponds in some ways quite well with the data. To the west of it there is a valley which is approximately 500 feet broad, and it does widen out considerably as it approaches the lake; thus a flat expanse is formed, well adapted for the site of a villa and a bath-building. It is true that the spurs which form the sides of it do not start directly from the Puy de Charmon, which in the Sauteyras theory must be the 'mons terrenus'. A plateau nearly a mile broad intervenes between the Puy and the head of the valley. This is certainly an objection from which none of the Aydat valleys is exempt, but perhaps in a writer like Sidonius it should not be pressed. But against Sauteyras two objections can be offered which seem conclusive: (a) There is no stream at all running down the Sauteyras valley, at least at the present day, and though Dumas-Damon says that such a stream exists,[3] he does not mark it on his map.

---

[1] Cf. 'Nouveaux Écl.', p. 277. [2] Rev. d'Auvergne, 1895, p. 331.
[3] l. c., p. 340.

(*b*) If the Sauteyras valley is the valley described by Sidonius there can be no doubt where the villa should have been. It must be looked for in the fields to the north and north-north-west of the farmhouse on either side of the modern road. In May 1932 I was kindly permitted to walk over these fields, which were partly arable and partly under grass. No sign of foundations could be seen, and a few sherds of pottery which were certainly post-Roman were all that could be picked up from the soil. Moreover, I was told that no remains had ever been found in these fields. Dumas-Damon would seem to imply that the site has been buried under the buildings of Sauteyras, but, as Crégut points out,[1] Sauteyras is situated on one of the spurs which bound the valley, and this is not the situation of Avitacum.

2. Crégut has argued for a site now occupied by the village of Aydat; and this view seems now to be generally accepted. Certainly a visitor sitting on the hill above the village might almost believe that he had found Avitacum. The valley opening out at its base and the running stream are in perfect harmony with the language of Sidonius. Moreover, Roman pottery and remains of an aqueduct have been found there.[2]

Nevertheless, against the village of Aydat certain objections can be raised. The village itself is situated not on the west side but at the south-west corner, and on Crégut's view, if the portico faced out over the lake, it must have been facing not east but at nearest north-north-east by north. Sidonius' narrative must clearly imply to any one who reads it through that the 'fluvius' which supplied his bath-building was not the same as the main affluent of the lake, yet on Crégut's view they are identical. Moreover, Sidonius tells us that the stream enters the lake at a place where it is un-approachable by reason of the mud. Yet this point, according to Crégut's theory, was not only within a stone's throw from the villa itself, but was identical with the meadow that

[1] 'Nouveaux Écl.', p. 254.
[2] 'Avitacum', pp. 97–9; 'Nouveaux Écl.', pp. 281–5.

C C

intervened between the villa and the lake. Sidonius, in fact, was playing tennis with Ecdicius on ground which in another passage he declares to be an impassable morass.

Furthermore, if the valley through which the stream flows was indeed the 'vallis' of Sidonius, the situation of the 'mons terrenus' creates a new difficulty. The only mountain to the west must be the Puy de la Rodde, as Crégut sees.[1] Yet the spurs bounding the valley not only do not issue from the Puy de la Rodde but have no connexion with it at all. In fact, in order to reach the 'mons terrenus' from the southern 'collis' it is necessary to cross the valley itself. This is utterly opposed to Sidonius' language.

It is true that Aydat village has what no other site can offer, Roman remains. Yet if the existence of the pottery shows that there was a Roman site at Aydat village it does not show that it was Avitacum, and as for the aqueduct, it was never planned or drawn, and in these circumstances we cannot be sure whether it really was either an aqueduct or of Roman date.

3. There remains the third valley, that between Sauteyras and Aydat village. A small brook flows down it, and though in its relation with the 'mons terrenus' (in this instance the Puy de la Rodde) it offers the same objection as the Sauteyras valley, that objection is perhaps not decisive. But the valley after all is only a slight hollow in the hillside and hardly deserves the name of a valley; and difficulties appear when we try to place the villa in relation to the stream. It is necessary to assume that the villa stood not on the edge of the lake, but on a little cliff some fifty feet above it, a cliff in which the brook has carved itself a steep ravine. This is scarcely consistent with the passage in which Sidonius says that the villa is bathed in the spray of the lake. Furthermore, as at Sauteyras, we can on this theory plot the position of the villa, and an examination of the ground produced again a few sherds, but nothing distinctively Roman.

[1] 'Avitacum', p. 77.

What then is to be done? The external evidence for Aydat is so strong that one hesitates to abandon it except for a site corresponding accurately with the data of Sidonius, and giving evidence of Roman habitation. And no such site has been proposed as yet, while at Aydat any site proposed seems to have some objection which is fatal to it. In fact, if a visitor well acquainted with Sidonius' work and not informed of the external evidence were taken to Aydat, I think that he would say: 'You have made a mistake, this is not the lake that Sidonius was describing, his villa was not situated upon its shore.' I can only offer the lame suggestion that the cliff at the base of the valley in site 3 has been formed by an earth-fall which has buried the villa out of all view. Such earth-falls are of frequent occurrence in the district, and an examination of the plateau of Gergovie shows clearly how they make difficult the interpretation of a historic site. It may be said that it is insulting historical method if we postulate a natural catastrophe for which no evidence exists in order to solve our problem. Yet desperate situations demand desperate remedies.

APPENDIX C

APOLLINARIS OF VOROCINGUM AND APOLLINARIS OF VAISON

TILLEMONT (*Mém. Ecclés.* xvi, p. 230) inferred that all the passages mentioning an Apollinaris other than Sidonius' son (i.e. *C.* xxiv. 52–3; *E.* ii. 9; iv. 4; iv. 12; v. 3; v. 6) were to be referred to the same person; and I believe that he is right. For (i) there is no mention of Apollinaris of Vaison in any letter which can be placed before Sidonius' episcopate. (ii) There is no mention of Apollinaris of Vorocingum in any letter which can be placed after it. (iii) In none of the letters addressed to Simplicius of Vaison is it stated that Sidonius had become bishop at the time of writing (*E.* iii. 11; iv. 7; v. 4). (iv) Both the letters addressed jointly to Simplicius and Apollinaris (*E.* iv. 4; iv. 12) must necessarily be placed after his election. (v) A certain Thaumastus of 'Tres Villae' (near Narbonne) appears in *C.* xxiv. 84–6, and is never heard of again, unless he is the *Thaumastus*, brother of Apollinaris of Vaison (*E.* v. 6. 1; v. 7. 1), who has never been heard of before. Individually these arguments have little weight, but cumulatively they are strong and justify Tillemont's conclusion.

APPENDIX D

THE PRAETORIAN PREFECTS OF GAUL (469–77)

THE chronology of the praetorian prefects at this time is very obscure. Between 469 and 477 we have to fit in Magnus Felix (Gennadius, *de Viris Illustr.*, 8; *E.* ii. 3. 1), Eutropius (*E.* iii. 6. 2), Protadius (*Edictum Glycerii* in Haenel, *Corpus Legum*, p. 260; see Borghesi, *Œuvres*, x, p. 747), and Polemius (*E.* iv. 14. 1). Of these Protadius, whose term is assigned to 473 by the edict (issued April 29th), is the only one to whom a date is definitely ascribed. If, as will be shown, Arles and Provence came into Visigothic hands in 473, and were not recovered by the Empire till 475 (*vide infra*, pp. 203, 209), there could have been no praetorian prefect in these years, and thus Borghesi's date of 474–5 for Felix's term will not do (x, p. 748). We are left with the years 469–72, 475–7 to fill. Now we find that *E.* ii. 3 occurs in a book in which none of the letters seem to belong to the period when Sidonius was bishop; we may then assume that Magnus Felix succeeded Arvandus in 469. The book, in which *E.* iii. 6 occurs, contains, with the exception of the concluding letter, none that seems to be subsequent to the capitulation of Clermont. In

E. iii. 6 there is not a hint that Clermont is being besieged, nor is there a word about fighting; for that reason it is plausible to assume that Eutropius succeeded Magnus Felix in 470 (so conjecturally, without argument, Borghesi, x, p. 745). Polemius presents still greater difficulty. He was praetorian prefect for at least two years (*E.* iv. 14. 1) after Sidonius had become bishop. We then have as possible dates either 471–2 or some date beyond 475 (so Borghesi, x, p. 750). Of these alternatives the former is far the more probable. The language of *E.* iv. 14 taken as a whole argues strongly against a view which would make Polemius prefect after Sidonius' captivity. And it is barely credible that such a phrase as '. . . quod te praefectum praetorio Galliarum . . . nostro affectu gaudemus, qui, si Romanarum rerum sineret aduersitas, aegre toleraremus, nisi singulae personae, non dicam prouinciae, uariis per te beneficiis amplificarentur', with its first persons plural, could be used at a time when Auvergne had become a Visigothic province and Sidonius a Visigothic subject. For these reasons, I reconstruct the succession of praetorian prefects as follows: 469 Magnus Felix, 470 Eutropius, 471–2 Polemius, 473 Protadius (until the capture of Arles by Euric after April 29th).

APPENDIX E

THE CHRONOLOGY OF THE SIEGES OF CLERMONT

THAT the chronology of the first years of Sidonius' bishopric would present great, perhaps insuperable, difficulties might have been foreseen by any one who knew that in his Collected Letters Sidonius was little concerned with chronological arrangement. Nevertheless, no serious attempt has been made to see whether the problems presented are capable of solution at all. Editors and biographers of Sidonius have for the most part neglected the hints offered by the chroniclers, and general historians have not, as a rule, possessed the necessarily minute knowledge of the letters. As, therefore, previous writers (with the honourable exceptions indeed of Fauriel and Stein) have contributed little or nothing of value,[1] their conclusions, being unsupported by argument, may for the most part be neglected.

Discussion of the chronology of the sieges is easiest, or rather only feasible, if we work backwards from a known date. The cession of Auvergne occurred, as is well known, after certain embassies in the reign of Nepos (474–5). Two of these embassies, at least, are known to us—that of the quaestor Licinianus and that of Epiphanius, bishop of Ticinum. The former we can date fairly closely. Sidonius writes,[2] almost certainly from Lyons, that Julius Nepos has sent Licinianus with a letter conferring the patriciate on Ecdicius[3]; and Licinianus' arrival at the city (probably Narbo) near which Magnus Felix was living is dated to the winter ('ninguidus dies'), at a time when the barbarians were expected to retire into winter quarters.[4] Thus Licinianus' embassy is quite securely dated to the winter of 474.[5] Now, according to

---

[1] Dom Vaissette (*Hist. de Languedoc*, Toulouse, 1872, vol. i, pp. 489–99) attempts a detailed chronology of the siege-operations. But his results are quite arbitrary, and would ill repay examination.

[2] *E.* v. 16. As it was written from a place where Sidonius' daughter, sisters, and mother were staying, and where his wife was not, and as Sidonius' family certainly came from Lyons (Coville, pp. 36–8 and p. 70), this view can be adopted with some confidence; and the language of the letter rather suggests that Sidonius was near the road into Gaul over the Alps when he received the news of Licinianus' mission. To Baynes's view (expressed in a letter to me) that *E.* v. 16 was written from the town of Clermont I cannot subscribe.

[3] Ib., § 1. Cf. Sundwall, p. 17.

[4] *E.* iii. 7. 4.

[5] Sirmond (p. 62) holds with very little probability that there were two embassies of Licinianus. Sundwall (p. 18) thinks that Licinianus made a peace which Euric broke on hearing that Ecdicius had been made *Magister militum praesentalis*. This is only a guess; but that he was not, at least, finally successful is obvious. Schmidt (i, p. 265) has no right to connect *E.* v. 12. 2 with Licinianus' embassy.

Ennodius, the mission of Epiphanius was successful:[1] we are therefore entitled to conclude that it was posterior in time to that of Licinianus. Otherwise we must hold with Sirmond[2] that after Epiphanius' embassy, Euric broke the treaty then made, and that Licinianus was sent subsequently. But for these events the period between Nepos' accession (June 25th) and the winter of the same year hardly allows enough time. For the dating of Epiphanius' embassy we have a curious but, I think, decisive *argumentum a silentio*. In a section of the Life describing the difficulties of the journey across the Alps nothing whatever is said of cold or snow, indeed there are references to 'cespes uiridans' and 'nemorea frons'.[3] But if the saint had crossed the Alps in the winter, Ennodius would certainly have described the hardships of his journey; and hence we can infer that the mission of Epiphanius is to be dated to late spring 475 at the earliest.[4] Much later than May 475 it cannot have been, for Ennodius, after making a very leisurely journey home,[5] found Nepos still emperor on his return—that is, before August 28th, 475.[6] Thus the cession of Auvergne and with it the writing of *E.* vii. 7 are dated to *c.* May 475. With this dating concurs the narrative of Jordanes, who states that after the cession of Auvergne Ecdicius retired to 'a safer place in Gaul'. From thence he was summoned to Rome by Nepos, and was superseded as *magister militum* by Orestes, who subsequently dethroned the emperor.[7] All these events must have taken time, and thus we need hardly look for a Gothic siege of Clermont in 475.[8] We now go back to 474. Sidonius' movements in that year are curious, but well documented and certain. 'In

---

[1] *Vita Epiphanii*, 91.   [2] p. 38.   [3] *Vita Epiphanii*, 83–4.

[4] Jullian (*Hist. de la Gaule*, vol. v, p. 159, n. 7) alleges that Epiphanius crossed the Alps in March. But the text of Ennodius gives no such precise date.

[5] Ib. 92–4.   [6] *Fast. Vind. Priores*, 616; *Auct. Haun.*, i, pp. 306–9.

[7] Jord., *Get.*, xlv. 241.

[8] Chaix alone (ii, pp. 166–77) assumes a siege in 475; he also dates the institutions of Rogations to the winter of 474–5, and this, as we shall see, is not possible.

the beginning of autumn,' he writes,[1] 'I was able to make a journey to Vienne'; and in the same letter he mentions an accusation that his cousin Apollinaris was trying to win Vaison for the 'nouus princeps'. This can be none other than Nepos,[2] and thus the letter is dated securely to *c.* September 474. At this time, he tells us, the fears of the Arvernians were in some degree mitigated owing to the season of the year.[3] The letter to Papianilla announcing that Ecdicius has been made patrician[4] is, as has been explained, probably from Lyons; but by the winter Sidonius was back at Clermont: at this time the Goths were expected to be going into winter-quarters, but were still in the vicinity of the town.[5]

While Sidonius was at Lyons, he saw[6] the carts full of grain collected by Patiens for the destitute population of south Gaul. One is naturally tempted to refer this letter to the autumn visit of 474, for otherwise we should have to assume a second journey to Lyons during the siege years. There is nothing inherently improbable in this: yet one is entitled to point out that no letters mentioning a journey taken during the siege exist, except the datable group of letters already mentioned (*E.* v. 6, 7, 16) and this letter undated. There is a presumption that they all date from the same year; and this presumption, though not logically conclusive, practically has

---

[1] *E.* v. 6. 1.

[2] Olybrius and Glycerius are ruled out for they were both Burgundian nominees. Binding (*Geschichte des burgund.-römischen Königreichs*, p. 81) made the barely intelligible suggestion that the 'nouus princeps' was Euric. According to Fauriel (i, p. 320), who dates *E.* v. 6 and 7 to 473, the 'nouus princeps' 'était sans doute Gondebaud'. This view is implicitly refuted by Coville (pp. 161–5) who should not, however, have said (p. 136) that Sidonius never calls a barbarian king 'princeps', for see *C.* vii. 219. Baret (p. 136) strangely assigns the letter to 470.

[3] *E.* v. 6. 1, 'cum Aruernorum timor potuit aliquantisper ratione temporis temperari'.

[4] *E.* v. 16. 1. No assumption of a remission of the siege should be drawn from the words 'uicinae obsidionis terror' of ib. 3. Dalton's 'imminent dread of siege' (ii, p. 70) is probably a mistranslation. It is more likely that 'uicinae' refers to place than to time.

[5] *E.* iii. 7. 4.   [6] *E.* vi. 12. 5, 'uidimus'.

weight. It is unfortunate that no certain inference of date can be drawn from the list of devastated towns mentioned in this letter: they would suit any date between 471 and 475.[1] If we place this letter, then, to winter 474–5, we have evidence for an incursion accompanied by the burning of crops in 474.

In order to construct a framework for the very difficult years 471–4, we have to answer certain questions: (i) What were the movements of Ecdicius in this period? (ii) What was the date of the mission of Constantius mentioned in *E.* iii. 2? (iii) When were the 'Rogations' of Mamertus introduced? (iv) What was the date of the mission of Avitus mentioned in *E.* iii. 1? Each of these questions will be treated in turn.

(I) If *E.* vi. 12 refers, as I have tried to show, to the autumn of 474, then we have the admittedly weak evidence of Gregory of Tours[2] that Ecdicius was in Burgundy in this year. If so, seeing that he was very probably at the Burgundian court at the time of the dispatch of *E.* iii. 3,[3] it might be plausible to date that important letter to 474. But there are difficulties. If the letter was written after Ecdicius had been made patrician (i.e. in winter 474–5), it is surprising that among the claims of Ecdicius to distinction that are mentioned in it the patriciate[4] does not occur; if, on the other hand, it was written before (i.e. summer 474), it is equally surprising that, while several reasons are given for Ecdicius' return, it is not hinted that it is desired because the Goths were besieging the town; yet we know that they were.[5] *Argumenta a silentio* are dangerous tools, yet on the evidence, such as it is, it seems preferable to assign this letter to a date

---

[1] Stein, indeed (p. 580, n. 3), seems to date this letter to 471. But the tone of the references to Clermont suggests a later date.

[2] Greg. Tur., *Hist. Franc.*, ii. 16 (24). Dalton (*Gregory of Tours*, ii, p. 496) rashly assumes that Burgundy is an error of Gregory's for Auvergne.

[3] See *E.* iii. 3. 9.

[4] We cannot safely decide from *E.* v. 16 whether Ecdicius was at that time in Burgundy or not.

[5] See *E.* v. 6. 1.

not later than winter 473–4. Nevertheless, even when we have dated the letter, we are not much nearer dating the events described in it. They may be referred with almost equal probability to the years 471, 472, or 473, though the language of the letter seems rather to suggest that some time had elapsed between the exploits and the letter describing them. We know that Ecdicius performed some feat in recompense for which he was promised the patriciate by Anthemius. This promise is hardly likely to have been given for services performed in 472, when Anthemius had enough to do to defend his own life and would hardly have troubled about the sieges of Clermont. We find then that Ecdicius performed some feat of arms in a year prior to 472; and this exploit of Ecdicius in the siege of Clermont is the only one which is reported to us. As I shall show that the sieges of Clermont began in 471, it seems plausible to combine the data of these two letters and to assign the events of E. iii. 3 to 471.[1] This view has been adopted in the text, as having more to commend it than any other, but certainty cannot be reached.

(II) It is not difficult to see close verbal resemblances between E. iii. 3 and E. iii. 2:[2] they suggest a man writing two letters by the same post. If that were so we could date Constantius' mission to 473–4.[3] On the other hand, we know that Sidonius revised his letters before publication, and certainly E. iii. 3 has the appearance of having been worked over. Thus a view based on such verbal correspondence is dangerous. This mission of Constantius occurred in winter;[4] but in what winter cannot be safely decided.

(III) The 'Rogations' of Mamertus are mentioned in two letters—E. v. 14 and E. vii. 1—and the second contains evidence from which we can elicit a date. It mentions the threat of an invasion by the Goths, that is, the letter is to be

<hr />

[1] So alone Seeck, *Untergang*, vi, p. 376.

[2] Cf. E. iii. 3. 3, 'semirutis murorum aggeribus' and 'ab omni ordine sexu aetate', with E. iii. 2. 1, 'semirutis moenibus' and 'omnis aetas ordo sexus'.

[3] So Tillemont (*Mém. Ecclés.*, xvi, p. 247).   [4] E. iii. 2. 3.

placed at the beginning of a campaigning season. In the same section Sidonius goes on to say: 'quod necdum terminos suos ab Oceano in Rhodanum Ligeris alueo limitauerunt, solam sub ope Christi moram de nostra tantum obice patiuntur'. At this time, then, it is obvious that the Goths were not in possession of Provence. Yet there can be no doubt that they were in possession of the whole or most of it by the end of 473. For we learn from *Chronica Caesaraugustana*[1] that in 473 Arles and Marseilles were captured by the Visigoths, and from *Chronica Gallica, a. dxi*,[2] that in the same year one of Euric's generals invaded Italy. When two chroniclers relate two different but absolutely consistent events to the same year they furnish mutual corroboration and each lends weight to the entry of the other. The 'Rogations' cannot, then, have been instituted later than the spring of 473. Nor can their date well be earlier than autumn 472, for the 'semper' of *E.* vii. 1. 1 argues not less than two previous 'irruptiones'—that is, in 471 and 472; and, as we shall see, if we assume an 'irruptio' in 470, there is not enough room for those letters in the early part of Sidonius' episcopate which convey no hint of incursions. Thus the institution of the 'Rogations' is fixed to 472–3,[3] and the chronology of the sieges is to be worked out in the following way: 471—an 'irruptio' defeated perhaps by Ecdicius; 472—an 'irruptio' in which the town was damaged by fire;[4] 473—a regular siege; 474—a siege accompanied by the burning of the crops and famine.[5]

[1] ad ann. 473 (ii, p. 222): 'his coss. Arelatum et Massilia a Gotthis occupata sunt'.

[2] 653 (i, p. 665): 'Vincentius uero ab Eurico quasi magister militum missus ab Alla et Sindila comitibus Italiae occiditur'. Stein arbitrarily ('Meines Erachtens', p. 585, n. 2) moves this invasion forward to 474, Schmidt equally arbitrarily dates it to 477 (i, p. 267).

[3] The institution of these 'Rogations' is usually placed in 474 or 475. Fauriel alone (i, p. 327) dates them correctly.

[4] For in 472–3 it had already been 'ambusta' (*E.* vii. 1. 2).

[5] That the incursions began in 471 was seen by Fauriel (i, p. 324), who is followed more or less by Dahn (v, p. 92, n. 2). According to Yver (pp. 32–3) Jordanes states that the attacks on Clermont began after the death of Anthemius (*Get.*, xlv. 238). Jordanes does not.

(IV) The early 'irruptiones' are illuminated by *E*. iii. 1; here the important words are 'illi (Gothi sc.) ueterum finium limitibus effractis . . . metas in Ligerim Rhodanumque proterminant'.[1] It is not hard to connect these words with the entry of *Chronica Gallica, a. dxi*, describing Gothic ravages around Arles in 471. Indeed, the letter cannot well be assigned to a later date, for it is hard to believe that the Goths would have agreed to abandon *Narbonensis* at any time after the defeat of the imperial army in 471. Thus the mission of Avitus must be dated to 471, and, as has been postulated in the text, it may well be closely connected with the Roman military expedition to south Gaul mentioned by the chronicler.[2] In that case 'quietiora' in *E*. iii. 1. 5 gives a further confirmation for the view that there was a Gothic 'irruptio' into Auvergne in 471.[3] References to this first Gothic 'irruptio' are also to be found in *E*. iv. 4 and iv. 6. The latter is written in the winter[4] and dissuades the family of Apollinaris from making a journey 'to the tomb of the Martyr', who is plausibly identified by Chaix[5] with St. Julian of Brioude. In *E*. iv. 4, which precedes it in time, Sidonius expresses the hope that he may visit Lyons—'si per statum publicum liceat': and again—'si non, quod etiam nunc ueremur, uis maior disposita confundat'.[6] If Chaix's conjecture is right—and it is, to say the least, extremely probable—the letter cannot be assigned to a date later than winter 471–2; for no one would surely have even considered taking his ladies on a journey to

---

[1] *E*. iii. 1. 5.

[2] *Chron. Gall., a. dxi*, 649 (i, p. 664), 'Antimolus ... cum Thorisario, Euerdingo, et Hermiano com. stabuli Arelate directus est: quibus rex Euricus trans Rhodanum occurrit, occisisque ducibus omnia uastauit.' I cannot agree with Schmidt (i, p. 265), who connects this embassy with that of Licinianus. Dahn (v, p. 83, n. 2) strangely identifies this Avitus with the emperor.

[3] To this incursion are probably to be assigned the circumstances of *E*. ix 9. 6, and in this letter a hint is given of the length of time (two months, § 7) during which the raid made travelling dangerous.

[4] *E*. iv. 6. 4, 'in maximo hiemis accentu'.

[5] ii, p. 116. Cf. *E*. vii. 1. 7; *C*. xxiv. 16–19.

[6] *E*. iv. 4. 2.

Auvergne in 472 if, as has been shown, there was an enemy burning the walls of Clermont. *E*. iv. 6 taken alone could be assigned to the winter either of 470–1 or of 471–2; but in the former alternative *E*. iv. 4 would have to be dated to 470, a year in which no Gothic incursion into Auvergne is recorded. 471, then, for *E*. iv. 4, and winter 471–2 for *E*. iv. 6 are the most probable dates.

It is unfortunate that we cannot date the truce mentioned in *E*. v. 12. The letter refers to a siege; and the words 'non per foederum ueritatem' seem to exclude the mission of Licinianus, who, as an accredited envoy of the empire, would have been responsible for making a *foedus*. But more we cannot say: the hopes of a truce may have existed at any time between 471 and 474.[1]

No more can we say definitely to what year the letters mentioning Burgundian assistance are to be dated. Of these there are two (*E*. iii. 4. 1 and vii. 11. 1). We can only say that the *inuidia* mentioned in both the passages does not well suit the period when Olybrius and Glycerius were on the throne.[2]

To the scheme here constituted there is apparently one serious objection. The chronology of the sieges as put forward includes an incursion in 471 and a serious siege in 472. Yet it is generally held that Sidonius became bishop of Clermont in 472, or at least not earlier than in the winter of 471.[3] The recognized chronology, however, which involved

[1] It is curious that the three editors, Dalton (ii, p. 64), Chaix (ii, p. 148), and Baret (p. 137)—who actually says 'date certaine'—all agree in the date which they assign for this truce. Such unanimity is rare, and it is the more curious that they should adopt the date against which there *is* something to be said—474. Chaix (ii, p. 145), indeed, by assigning an arbitrary date to *E*. v. 20 assumes a truce made through the praetorian praefect at Arles and soon broken. This is most unlikely.

[2] Fertig (ii, p. 10) and Coville (p. 179) are inclined, rather hypercritically, to doubt whether the Burgundians actually sent a force to Auvergne, but the words 'propugnantum inuidia' of *E*. iii. 4. 1 are fairly decisive.

[3] The various opinions of scholars on this point are collected by Coville (p. 68, n. 5).

assigning the election at Bourges to 472 and crowding all the episcopal letters not referring to a siege into that year, must have been felt to have been exceedingly cramping: on the chronology of the sieges as here put forward it is, of course, impossible. The recognized date of Sidonius' election is based on a complicated argument of chronology. In *E*. vi. 1. 3, a letter written, as is supposed, shortly after the election, it is stated that Lupus of Troyes has been bishop for forty-five years; according to *Vita Lupi*, 4, Lupus became bishop two years before he made his journey to Britain with St. Germanus; and this journey is dated by Prosper Tiro to 429 (*Chron.* 1301 (i, p. 472)). From these facts, it is alleged, the date of *E*. vi. 1 can be fixed to 472. If they are sound, it can indeed; but the argument is such that, if doubt can be cast upon any one of the stages of it, the validity of the whole is impugned. And doubt can be cast upon more than one of them. In the first place, it is quite impossible to determine what interval of time there is between the election of Sidonius and the dispatch of *E*. vi. 1: in the second place, Sidonius' dealings with figures where he can be checked[1] show that only a qualified reliance should be placed upon his statements where he cannot: in the third place, Krusch has proved independently[2] that the chronology of the *Vita Lupi*, especially in regard to the election of Lupus as bishop, is not admissible. Thus with doubt cast upon three of its stages the argument drawn from *E*. vi. 1. 3 must be abandoned. On the new view of the chronology here put forward it is necessary to assign the election at Bourges to a date prior to the first incursion of the Visigoths into Auvergne in 471: and it is hardly probable that it is to be dated to 469, the year of the battle of Bourg-de-Déols, when Euric's troops were conquering Berry, for there is no indication that the election was held in the midst of strife. 470 we may assume then as the date of it. Sidonius had not very long been a bishop at the time of the

---

[1] See *E*. ix. 16. 3, vv. 21–32.
[2] In *M.G.H.*, *script. rer. Merou.*, iii, p. 118.

Bourges election,[1] so he must have been himself elected in 470 or 469. To decide between these years is not easy. In view, however, of the fact that there are some thirty letters of Sidonius (one of which mentions a severe illness) which appear to have been written between the time of his election and the beginning of the incursions, we should do better to date the election to 469, so as to provide the maximum interval of time for these letters. And if, as I have tried to show in the text, the Visigoths were not in occupation of the Rouergue at the time when *E.* vi. 15 was written, that gives another hint that 469 is the right date. If that is the year, the date must be some time after June 28th (*C.* xvii. 1), but before the winter (*E.* iv. 15. 3).[2]

APPENDIX F

THE PEACE OF 475

We know that in the late spring of 475 Epiphanius, bishop of Ticinum, was sent by Nepos on an embassy to Euric to make peace.[3] Epiphanius demanded 'ut reductis ad fidem mentibus terrae sibi conuenae dilectionis iure socientur'.[4] Euric, persuaded by the bishop's eloquence, agreed; and Epiphanius returned home—'inito pactionis uinculo'.[5] We have also, however, two letters of Sidonius referring to a treaty made by Gallic bishops. In one of them (*E.* vii. 6) he mentions that the bishops Basilius of Aix, Leontius of Arles, Faustus of Riez, and Graecus of Marseilles, are responsible for the treaties between the two kingdoms;[6] he urges them to obtain

---

[1] 'nouitatem', *E.* vii. 5. 2.

[2] 469 or 470 is the date adopted by Mommsen (*ap.* Lütjohann, p. xlviii); *c.* 470 that of Schmidt (i, p. 264).

[3] Ennodius, *Vita Epiphanii*, 81–94.   [4] Ib. 88.   [5] Ib. 91.

[6] 'pacta utriusque regni', *E.* vii. 6. 10. Dahn (v, p. 91, n. 3) and Schmidt (i, p. 265, n. 3) maintain that the 'kingdoms' ('regna') must refer to the Visigoths and the Burgundians, because Sidonius never calls the empire 'regnum'. It is quite possible that their conclusion is correct, but their reason

for the bishops the right of ordination in the parts of Gaul which the treaty will assign to the Goths. In the other, addressed to Bishop Graecus (*E.* vii. 7), he blames him bitterly for having concluded a treaty which has sacrificed Auvergne. That of these two letters *E.* vii. 6 is the earlier may be accepted as reasonably certain; for in it we see that Sidonius does not yet know that Auvergne is to be abandoned, whereas in *E.* vii. 7 he quite obviously does. Neither of these letters, however, contains any mention of Epiphanius: and, conversely, there is no mention in Ennodius' Life of the 'Treaty of the Four Bishops'. The relation between them, therefore, is not an easy matter to discover, and has been differently expressed by different historians. Schmidt[1] holds that there was an attempted negotiation of the Four Bishops which failed, and that after it the embassy of Epiphanius succeeded in making terms.[2] Such a conclusion has the merit of resting on the clearly indicated success of Epiphanius' mission; yet it has to meet Dalton's unanswerable query: 'If the embassy of the Four Bishops did not succeed, why was Sidonius so indignant with Graecus?'[3] It is necessary, in fact, to adopt the conclusion of Hodgkin[4] and Allard,[5] that there was an embassy of the Four Bishops who merely ratified the convention made by Epiphanius, and perhaps drew up specific terms for it. And this too is implied in a rather obscure sentence in the letter to Graecus.[6] 'Vobis primum pax', he writes, 'quamquam principe absente non solum tractata reseratur, uerum etiam tractanda committitur.' Here the sense of the words

is not valid. It is true that in his prose Sidonius never seems so to call the empire; but he does so several times in his poems (see *C.* ii. 21; v. 5; v. 11, &c.), and it is called 'regnum' by Majorian (*Novell.*, ii. 1).

[1] i, p. 265.

[2] This is also the view substantially of Dahn (v, p. 95) and Yver (pp. 35–6).

[3] i, p. xlii, n. 3. Fertig, indeed (ii, pp. 17–18), holds that it did succeed, but that Ecdicius succeeded in preserving the city with Burgundian help, and that it was only finally surrendered by the treaty of Epiphanius. But Fertig's narrative crowds too many events into the summer of 475 as well as being inconsistent with Jordanes, *Get.*, xlv. 241.

[4] ii, p. 504, n. 1.   [5] pp. 158–60.   [6] *E.* vii. 7. 4.

seems to be that Graecus is responsible for drafting the terms of a treaty: it is not at all implied that Graecus is the originator of a peace, it is rather implied that he is not. With this difficulty solved, however, we have a much harder one to face. We must explain why Epiphanius' mission was regarded as successful if it involved the cession of large portions of the empire. Admittedly Ennodius' work is panegyric, but even for a panegyric this is somewhat unexpected.

Yet an answer can be found if we look carefully into the history of the period. The entries in the *Chronica Gallica* and *Caesaraugustana*[1] show us that in 473 and 474 Provence was in Visigothic hands. On the other hand there is the clearest evidence that it was conquered by Euric from Odoacer in 476 or 477.[2] How could they win it, if they had already won it in 473? It is not sound criticism to assume that both the chronicle entries are misdated; for they are completely independent. We find then that at some time between 474 and 477 the Goths lost possession of Provence: we also find that in 475 there was an embassy which, though responsible for the cession of Auvergne, was regarded in Italy as a success. Is it great hazard to connect the two? Nepos' statesmanship will not then be mere weakness and folly: by acquiescing in the loss of the districts which Euric had already annexed and surrendering the distant Auvergne over which he could now exercise no real control he gained the contiguous Provence with Arles and Marseilles. It was ingratitude certainly, but it was not weakness.[3]

Yet even if we can in this way understand why the 475

---

[1] *Chron. Gall.*, a. dxi, 653 (i, p. 665); *Chron. Caesaraugust.*, ii, p. 222; cf. Jord., *Get.*, lvi. 284.

[2] *Auct. Haun., Ordo Prior*, i, p. 309 (476); *Chron. Gall.*, a. dxi, 657 (i, p. 665) (477); Candidus, fr. 1 (*F.H.G.*, iv, p. 136); Procop., *Bell. Goth.*, i. 12. 20. *Vide supra*, p. 164, n. 6.

[3] I do not wish the argument to rely on the distorted passage of Jord., *Get.*, xlvii. 244. Nevertheless, unless we see in it a 'telescoping' of two accounts of the capture of Arles and Marseilles under two different dates, it is hard to explain why it appears in this form.

treaty was regarded from the Italian point of view as a gain, we may still wonder why Euric acquiesced in the loss of Provence. The answer is given by a very curious passage of Ennodius' Life:[1]

inter eum (Nepotem sc.) et . . . Getas . . . orta dissensio est, dum illi Italici fines imperii, quos trans Gallicanas Alpes porrexerat, nouitatem spernentes non desinerent incessare, e diuerso Nepos . . . districtius cuperet commissum sibi a deo regnandi terminum uindicare.

The general sense of this is clear. There was a definite attempt made by the empire to claim ('uindicare') the lands across the Alps. The passage, obscure and almost meaningless[2] unless we assume a Gothic conquest of Provence in 473, becomes at once full of point if we do. Whether any battle took place between the Goths and the imperial troops we cannot say, but the words put into Epiphanius' mouth—'qui licet certamina non formidet, concordiam prius exoptat'[3]—suggest that the troops were rather used for an armed demonstration. They were not the whole of Nepos' army,[4] and it is probable that he did not himself accompany them.[5]

If this reconstruction be adopted, it explains why Nepos was able to make what was really a very favourable treaty with Euric—and it does more. It gives point to the praises of Nepos uttered by Sidonius—'Iulius Nepos, armis summus Augustus'[6]—and it gives a real motive for the supposed intrigues of Apollinaris at Vaison.[7] If an imperial army had

---

[1] *Vita Epiphanii*, 80; cf. Hodgkin, ii, p. 502, n. 1.

[2] Dahn (v, p. 96, n. 5) thinks that in this speech Ennodius was transferring the conditions of Theodoric's time back into that of Nepos—a very far-fetched defence, but at least preferable to Büdinger's desperate theory that the words 'quos . . . porrexerat' are spurious (p. 954). Historical criticism would be easy if we could always lay the blame upon the scribes.

[3] *Vita Epiphanii*, 88.

[4] Cf. *Auct. Haun.* (i, p. 307, 309), 'Orestes Roma mittitur cum robore exercitus'; and *Anon. Val.*, vii. 36.

[5] See *E.* v. 16. 1, cf. *Anon. Val.*, vii. 36; *E.* vii. 7. 4.

[6] *E.* v. 16. 2.

[7] *E.* v. 6. 2 and 7. 1.

actually crossed the Alps,[1] his action is seen to be no mere leap in the dark, but an act of sane policy, part perhaps of a considerable plan.[2]

APPENDIX G

ON THE DATE OF SIDONIUS' DEATH

THERE is no evidence from which the date of Sidonius' death can be fixed with any certainty. For a *terminus post quem* we have the date 479–80 assigned to *E.* ix. 12; and for a limit in the other direction we have the knowledge that Aprunculus, who succeeded Sidonius as bishop of Clermont, died in 490 (Greg. Tur., *Hist. Franc.*, iii. 2; Duchesne, *Fastes Épiscopaux*, ii, p. 35). But greater precision we cannot obtain, for we do not know how long Aprunculus was bishop of Clermont.

The Epitaph of Sidonius tells us that he died 'xii Kal. Septembres, Zenone imp.' Mommsen (ap. Lütjohann, p. xlix) emended 'imp.' into 'iterum consule' and thus dated Sidonius' death to 479. But, though this is accepted by such authorities as Stein (p. 546, n. 3) and Duchesne (l.c., ii, p. 34), it is entirely arbitrary, and the evidence of the Epitaph can be explained without resorting to conjectural emendation.

On p. 166, n. 2, I have given reasons for doubting whether the Epitaph of Sidonius has any independent authority and have shown its dependence on ps.-Gennadius. Now the author of the Epitaph could know the day of Sidonius' death,[3] because it was that on which his festival was celebrated, but his further statement, 'Zenone imp.', is nothing more than a

[1] Who the mysterious hostes of Jord., *Get.*, xlv. 241 may be is quite uncertain, but Hodgkin with some probability sees a reference to a projected attack on the Burgundians (i, p. 510). Sundwall (p. 18) seems to imagine that peace was concluded by Orestes after the death of Nepos: this contradicts Ennodius.

[2] We may notice the obscurely expressed feeling of Sidonius that the barbarians may not always be at Lyons (see *E.* v. 7. 7).

[3] The day given by the epitaph is that adopted by the church of Clermont: the Roman martyrologies give Aug. 23rd or Aug. 24th.

simple inference from the statement of ps.-Gennadius that Sidonius flourished under the reigns of Leo and Zeno.

Germain (pp. 181–2) notes that Gregory (*Hist. Franc.*, ii. 15 (23)) brings into close connexion with the death of Sidonius the moment 'when rumours of Frankish power echoed in Gaul'. From this he concluded that Sidonius survived to a date later than the battle of Soissons (*c.* 488). But examination of the text of Gregory must make it doubtful whether he actually intended such a synchronization and whether from the nature of his sources it would have been of any value if he had. We must be content with some unknown year in the decade 480–90.[1]

[1] Various theories as to the date of Sidonius' death are quoted by Coville (p. 73, n. 1).

BIBLIOGRAPHY

TEXTS OF SIDONIUS

1. C. Lütjohann, *Gai Sollii Apollinaris Sidonii Epistulae et Carmina*, in *M.G.H.*, *auct. ant.*, vol. viii, with additional introductory matter by Leo and Mommsen, list of parallel passages by Geisler, and indexes by Mommsen and Gruppe; Berlin, 1887.

   [Mommsen's introduction and index contain a complete and well-documented summary of Sidonius' life with a useful family tree; but his inferences are sometimes rash.]

2. P. Mohr, *C. Sollius Apollinaris Sidonius*; Leipzig (Teubner), 1895.

   [The critical editions of Lütjohann and Mohr, which differ little from each other, supersede all earlier texts.]

3. J. Sirmond, *C. Sol. Apollin. Sidonii Aruernorum Episcopi Opera*, ed. ii; Paris, 1652.

   [Sirmond's commentary is of the highest merit, and is still indispensable, especially for the Poems.]

4. E. Baret, *Œuvres de Sidoine Apollinaire*; Paris, 1879.

   [The biographical and historical introduction (pp. 1–106) is of no great importance, but the remarks upon the language of Sidonius (pp. 106–23) are very valuable. The chronological scheme of letters and poems (pp. 123–45) is amateurish.]

TRANSLATIONS

English.

5. O. M. Dalton, *The Letters of Sidonius*, two volumes; Oxford, 1915 (with full introduction and notes).

   [Well written and full of sound criticism. The most successful attempt at dating the letters.]

French.

6. J. F. Gregoire and F. Z. Collombet, *Œuvres de C. Sollius Apollinaris Sidonius*, three volumes; Paris, 1836.

   [The most important translation of the complete works. Uneven and apt to shirk difficulties.]

   [Translations of various passages in the Letters and Poems are also to be found in most of the authors who deal with Sidonius. (See Hodgkin, Chaix, *infra*.)]

ANCIENT AUTHORITIES

7. Avitus, Bishop of Vienne (ob. *c.* 524), *Opera*, ed. Peiper, in *M.G.H.*, *auct. ant.*, vi; Berlin, 1883.

8. Candidus (fl. 490), *Fragmenta*, ed. Müller in *F.H.G.*, vol. iv, pp. 135–7; Paris, 1885.

9. Cassiodorus (*c.* 490–583), *Variae*, ed. Mommsen in *M.G.H.*, *auct. ant.*, xii; Berlin, 1894. (For Cassiodorus' *Chronicle* see below 11 *h.*) [Cassiod. *Var.*]

10. Claudianus Mamertus (fl. 470), *Opera*, ed. Engelbrecht in *C.S.E.L.*, vol. xi; Vienna, 1885. [Claud. Mam.]

11. *Chronica Minora*, ed. Mommsen, 3 vols. in *M.G.H.*, *auct. ant.*, vols. ix, xi, xiii; Berlin, 1892–8.

[All references to Chronicles in the text are to section and page of these volumes.]

(*a*) Anonymus Valesii, Pars Posterior. [Anon. Val.]
(*b*) *Fasti Vindobonenses Priores.* [*Fast. Vind. Priores.*] } *Consularia Italica,*
(*c*) *Continuatio Hauniensis Prosperi.* [*Add. Prosp. Haun. & Cont. Prosper.*] } i, pp.
(*d*) Prosper Tiro; i, pp. 341–499. } 252–339.
(*e*) *Chronica Gallica ann. cccclii & dxi;* i, pp. 615–66. [*Chron. Gall. a. cccclii & dxi.*]
(*f*) Hydatius, ii, pp. 1–36.
(*g*) Marcellinus Comes, ii, pp. 37–108.
(*h*) Cassiodorus, ii, pp. 109–61. [Cassiod., *Chron.*]
(*i*) Victor Tonnennensis, ii, pp. 184–206. [Victor Tonn.]
(*k*) *Chronica Caesaraugustana*, ii, pp. 221–3. [*Chron. Caesaraug.*]
(*l*) Marius of Aventicum, ii, pp. 225–39.
(*m*) Isidore, *Historia Gothorum*, ii, pp. 241–95.
(*n*) *Laterculus Imperatorum ad Justinum* i, iii, pp. 418–23.

12. *Epistolae Arelatenses*, ed. Dümmler, in *Epistulae Merouingici et Karolini aeui*, vol. i, in *M.G.H.*, *Epist.*, vol. iii; Berlin, 1892.

[See especially *Epp. Arelat.*, 15, 19, 20; *Epp. Austrasicae*, 23.]

13. Ennodius (fl. 500), *Opera*, ed. Vogel in *M.G.H. auct. ant.*, vii; Berlin, 1885.

14. *Anonymus Sidonii Epitaphius* in Cod. Matrit., Ee. 102 (*circa* saec. viii, *ut vid.*) printed in editions of Lütjohann (1) (pp. vi and xliv) and Mohr (2) (p. 388).

15. Evagrius (ob. *c.* 600), *Historia Ecclesiastica*, ed. Bidez and Parmentier; London, 1898.

16. Faustus, Bishop of Riez (fl. 469), *Epistulae*, ed. Krusch, in same volume with Lütjohann's *Sidonius*. (See 1.)

17. Gennadius (fl. *c.* 500), *De Viris Illustribus*, ed. Herding; Leipzig, 1879.

> [Ch. 92 contains a short life of Sidonius: it is absent from most manuscripts and is probably by a later hand, not later than saec. vii. It is of no great value. See Mommsen ap. Lütjohann, p. xliv.]

18. Gregory of Tours (fl. 575), *Opera*, in *M.G.H.*, *script. rer. Meroving.*, vol. i, *Historia Francorum*, ed. Arndt; *Scripta Minora*, ed. Krusch; Hanover, 1884, 1885. *Historia Francorum* alone, ed. Omont and Collon, revised by Poupardin; Paris, 1913. English translation of the *Historia Francorum* by O. M. Dalton, two volumes; Oxford, 1927. [Greg. Tur.]

19. John of Antioch (saec. vii), *Fragmenta*, ed. Müller in *F.H.G.*, vol. iv, pp. 535–622; Paris, 1885. [John Ant.]

20. John Malalas (fl. 560), *Chronographia*, book xiv, ed. Migne, in *Patr. Graec.*, vol. xcvii.

21. Jordanes, *Romana* and *Getica*, ed. Mommsen, in *M.G.H.*, *auct. ant.*, vol. v, part i; Berlin, 1882.

22. Paulinus Petrocordiae (fl. 480), *De Vita Sancti Martini*, ed. Petschenig in *Poet. Christ. Minores*, vol. i, *C.S.E.L.*, vol. xvi; Vienna, 1888.

23. Paulus Diaconus (fl. saec. viii), *Historia Miscella*, book xv, ed. Droysen, in *M.G.H.*, *auct. ant.*, vol. ii; Berlin, 1879.

24. Priscus (fl. *c.* 474), *Fragmenta*, ed. Müller, in *F.H.G.*, vol. iv, pp. 69–110; Paris, 1885.

25. Procopius (fl. 527–63), *Bellum Vandalicum* and *Bellum Gothicum*, ed. Haury; Leipzig, 1905.

26. Ps.-Fredegar (fl. *c.* 650), *Chronicon*, ed. Krusch, in *M.G.H.*, *script. rer. Meroving.*, vol. ii; Hanover, 1888.

27. Ruricius, *Epistulae*, ed. Krusch, in same volume with Lütjohann's *Sidonius*. (See 1.)

28. Salvian (*c.* 400–80), *de Gubernatione Dei*, ed. Halm, in *M.G.H.*, *auct. ant.*, vol. i, part i; Berlin, 1877.

29. Theophanes, *Chronographia*, ed. de Boor; Leipzig, 1923.

INSCRIPTIONS

30. Allmer and Dissard, *Musée de Lyon, Inscriptions Antiques*; Lyons, 1888–94.

31. *Corpus Inscriptionum Latinarum*; Berlin, 1862– . [*C.I.L.*]

32. De Rossi, *Inscriptiones Christianae Urbis Romae*; Rome, 1857–88
33. Dessau, *Inscriptiones Latinae Selectae*; Berlin, 1892–1916.
34. Le Blant, *Inscriptions chrétiennes de la Gaule*; Paris, 1856–92.

LAWS, ETC.

35. *Canones Apostolorum et Conciliorum Veterum*, ed. Bruns; Berlin, 1839.
36. *Codex Theodosianus, Novellae* and *Constitutiones Sirmondianae*, ed. Mommsen and Meyer; Berlin, 1905.
37. *Corpus Juris Civilis*, ed. Krüger, Mommsen, and Scholl; Berlin, 1928–9.
38. *Constitutio Honorii de vii Prouinciis*, Text in (1) Sirmond's edition of Sidonius (*vide supra* 3), (2) Haenel, *Corpus Legum*, pp. 238–9; Leipzig, 1857, (3) Dümmler, *Epist. Arelat.*, pp. 13–15 (*vide supra*, 12), (4) Carette, *Assemblées provinciales*, pp. 460–3 (*vide infra*, 61).
39. *Edictum Imp. Glycerii*, ed. Haenel, *Corpus Legum*, p. 260 and Migne, *Patr. Lat.*, vol. lvi, p. 898.
40. *Lex Romana Visigothorum*, ed Haenel; Leipzig, 1849.
41. *Leges Visigothorum*, ed. Zeumer, in *M.G.H.*, *Leges*, vol. i; Hanover and Leipzig, 1902.

MODERN WORKS DEALING PRINCIPALLY WITH SIDONIUS

Biographical.

42. P. Allard, *St. Sidoine Apollinaire*; Paris, 1910.

[Disappointing; neither critical nor accurate. Laudatory.]

43. Arnold, Article—'Sidonius'—in Herzog-Hauck's *Realencyklopädie für protestantische Theologie und Kirche*, vol. xviii, pp. 302–9; Leipzig, 1906.

[A most sober, accurate, and well-documented sketch of Sidonius. Indispensable.]

44. M. Büdinger, 'Apollinaris Sidonius als Politiker' in *Sitzungsberichte der Wiener Akademie*, xcvii (1881), pp. 915–53.

[Suggestive, but wild in chronology.]

45. L. A. Chaix, *Saint Sidoine Apollinaire et son siècle*; Clermont-Ferrand, 1866.

[A valuable compilation but quite uncritical. Very laudatory.]

46. A. Coville, *Recherches sur l'Histoire de Lyon (450–800)*, chs. i–iii; Paris, 1928.

[Thoughtful, though not in my opinion completely successful reconstruction of the *Coniuratio Marcelliana*. The most up-to-date work on the history of the Burgundian settlements. Very full and useful but ill-arranged bibliographical notes.]

47. L. Duval-Arnould, *Études d'Histoire du Droit romain au 5ᵉ siècle d'après les lettres et les poèmes de Sidoine Apollinaire*; Paris, 1888.

[Useful discussions of points of detail. His view of the position and aims of Avitus has been adopted in the text.]

48. A. Esmein, 'Sur quelques Lettres de Sidoine Apollinaire' in *Rev. gén. du Droit*, 1883, pp. 305 et seq. (Also published separately; Paris, 1885.)

[Juristic studies like no. 47. Especially useful for the legal questions arising from *E*. vi. 4.]

49. F. Fertig, *Sidonius Apollinaris und seine Zeit*, three pamphlets; Würzberg and Passau, 1845–8.

[Rather naïve and amateurish. The literary criticism (in no. iii) is good.]

50. A. Germain, *Essai littéraire et historique sur Apollinaris Sidonius*; Montpellier, 1840.

[Simple, straightforward, and sensible. The best specimen of the laudatory biography.]

51. A. Kaufmann, 'Sidonius Apollinaris' in *Neues schweizerisches Museum*, 1865 [*sic*, but issued subsequent to the publication of no. 45], pp. 1–28.

[A counterblast to Chaix's work. One-sided, but exceedingly able and acute exposure of Sidonius' weaknesses. Indispensable. Of his other works, *Die Werke des C. Sollius Apollinaris Sidonius als eine Quelle für die Geschichte seiner Zeit*, Göttingen Dissertation, 1864, was unfortunately inaccessible to me. 'Rhetorenschule und Klosterschule' in Raumer's *Historische Taschenbuch*, pp. 30–40, is less important.]

52. A. Klotz, Article 'Sidonius' in *Pauly-Wissowa*, New Series, ii, pp. 2230–8; Stuttgart, 1923. [Klotz in *P.-W.*]

[Inferior generally to Arnold's article (*vide supra*, 43). Useful for the literary history of the poems and letters.]

53. Mommsen, 'Apollinaris Sidonius und seine Zeit' (1885) in *Reden und Aufsätze*, pp. 132–42; Berlin, 1912.

[A very suggestive and interesting essay.]

54. H. Peter, *Der Brief in der römischen Litteratur*, Chapter V d., pp. 156–8; *Abhandl. der Königl. sächs. Gesell. der Wissenschaften, philol.-hist. Klasse*, xx, 3; Leipzig, 1901.

[Analyses closely the manner in which Sidonius composed the Epistles and prepared them for publication.]

55. L. S. Le Nain de Tillemont, *Mémoires Ecclésiastiques*, vol. xvi; Paris, 1712.

Stylistic.

56. R. Holland, *Studia Sidonia*; Leipzig, 1905. See also nos. 1, 4, 5, 51, and 53.

GENERAL WORKS ON THE HISTORY OF THE PERIOD

57. N. Baynes, articles in *J.R.S.*, vol. xii (1922), pp. 222–8; vol. xviii (1928), pp. 224–5; and *History*, vol. xiv, no. 56, pp. 290–8.

[Discusses the problems connected with the reigns of Avitus and Majorian, and criticizes Pirenne's view (*vide infra* no. 78) on the extent of Mediterranean trade in saec. v.]

58 W. Bright, *The Canons of the First Four General Councils*, second edition; Oxford, 1892.

59. J. B. Bury, *History of the Later Roman Empire*, vol. i; London, 1923.

60. *Cambridge Mediaeval History*, vol. i, articles by C. H. Turner, L. Schmidt, E. Barker, E. C. Butler, P. Vinogradoff, and H. F. Stewart; Cambridge, 1911.

[Full bibliographies. Not much about Sidonius.]

61. L. Cantarelli, 'Annali d'Italia (455–76)' in *Studi e Documenti di Storia e Diritto*, xvii (1896), pp. 39–123.

62. E. Carette, *Les Assemblées provinciales de la Gaule romaine*; Paris, 1895.

[A valuable discussion of the trial of Arvandus, pp. 333–54.]

63. E. Dahn, *Könige der Germanen*, vols. v, vi, and xi; Würzberg and Leipzig, 1866–1911.

[Meticulous. Better on internal organization than on political history.]

64. Sir S. Dill, *Roman Society in the Last Century of the Western Empire*, especially book ii, ch. iv; London, 1925.

65. C. Fauriel, *Histoire de la Gaule méridionale sous les Conquérants germains*, vol. i; Paris, 1836.

[Full of good sense and good judgement. Very far from obsolete.]

66. Fustel de Coulanges, *Histoire des Institutions politiques de l'ancienne France*; vol. ii, *les Invasions germaniques*; vol. iii, *la Monarchie franque*; vol. iv, *l'Alleu et le Domaine rurale*; Paris, 1924–7.

67. E. Gibbon, *Decline and Fall of the Roman Empire*, ed. Bury, vol. iv; London, 1925.

[*Bury-Gibbon.*]

68. T. Haarhof, *Schools of Gaul*; Oxford, 1920.

69. T. Hodgkin, *Italy and her Invaders*, vols. i and ii (especially book iii, chs. iii–vii); Oxford, 1931.

[Some errors; but full of sound sense, like all that Hodgkin wrote.]

70. C. Jullian, *De la Gaule à la France*; Paris, 1922.

[Stimulating, tendentious work; enunciates the 'separatist' theory of Avitus' attempt.]

71. G. Lardé, *Le Tribunal du Clerc dans l'Empire romain et la Gaule franque*; Moulins, 1920.

72. E. Lavisse, *Histoire de la France*, vol. ii, part i (By Bayet, Pfister, and Kleinclausz); Paris, 1911.

[Valuable general summary with good photographs.]

73. C. Lécrivain, *Le Sénat romain depuis Dioclétien à Rome et à Constantinople*; Paris, 1888.

74. A. Longnon, *Géographie de la Gaule au sixième siècle*; Paris, 1878.

75. E. Löning, *Geschichte des deutschen Kirchenrechts*, 2 vols.; Strassburg, 1878.

76. F. Lot, *La Fin du Monde antique et le Début du Moyen Âge*; Paris, 1927.

77. F. Lot, *L'Impôt foncier et la Capitation personnelle sous le Bas-Empire et à l'Époque franque*; Paris, 1928.

[Profound study of the development of the 'villa-system', especially in Gaul.]

78. H. Pirenne, *Les Villes du Moyen Âge*; Brussels, 1927.

79. M. Rostovtzeff, *Social and Economic History of the Roman Empire*; Oxford, 1926.

80. L. Schmidt, *Geschichte der deutschen Stämme*, vol. i; Berlin, 1910.

[Indispensable.]

81. O. Seeck, *Geschichte des Untergangs der Antiken Welt*, vol. vi; Stuttgart, 1920. [Seeck, *Untergang*.]

82. O. Seeck, *Regesten der Kaiser und Päpste*; Stuttgart, 1919. [Seeck, *Regesten*.]

[Valuable for chronology.]

83. O. Seeck, articles in *Pauly-Wissowa*: Avitus, Majorian, Libius Severus, Anthemius, Ecdicius, Euric; Stuttgart, 1894. [Seeck, in *P.-W.*]

[All Seeck's work is valuable, but must be cautiously used.]

84. E. Stein, *Geschichte des spät-römischen Reichs*, vol. i; Vienna, 1928.

[Contains some very sound observations on Sidonius. His reconstruction of the events in 456–8, however, goes astray, in my opinion.]

85. J. Sundwall, *Weströmische Studien*; Berlin, 1915.

[Brings forward many new theories for the history of Gaul 468–80, all of them ingenious, not all satisfactory. The fifth-century prosopography is indispensable.]

86. G. Tammasia, *Egidio e Siagrio*; Turin, 1886. (Extr. from *Rivista Storica Italiana*, vol. iii, fasc. ii.)

87. L. S. Le Nain de Tillemont, *Histoire des Empereurs*, vol. vi; Paris, 1738.

88. G. Yver, 'Euric Roi des Wisigoths', in *Études d'Histoire du Moyen Âge dédiées à Gabriel Monod*; Paris, 1896.

[Valuable on his own subject, unreliable outside it.]

INDEX

Abraham, monk of Clermont, 161.
Aegidius, 45 n.[4], 50, 89, 90.
Aëtius, Flavius, 20, 21, 24, 26, 33, 47.
Agathe [Agde], council of, 121.
Agathias, 16, 17.
Agricola, son of Eparchius Avitus (*vide infra*) 20 n.[2], 23, 69.
Agrigentum, 37.
Agrippinus, *comes*, 90.
Agrippinus, priest, 120.
Agroecius, bishop of Sens, 128.
Alamanni, 25, 27, 144.
Alani, 2, 58 n.[1], 99.
Alaric, 34.
Alcima, daughter of Sidonius, 84.
Alypia, daughter of Anthemius, 96.
Ambrose, 115, 118.
Anianus, bishop of Orléans, 168.
Anthemiolus [Antimolus], son of Anthemius (*vide infra*), 149, 204.
Anthemius, western emperor (A.D. 467–72), 5 n.[2], 59 n.[3], 94–101, 104, 107, 113, 139, 149, 150, 153, 202, 203 n.[5].
Antiquarianism, in Later Roman Empire, 15, 16.
Apollinaris, cousin of Sidonius, 64, 68, 140, 151, 157, 195, 196, 200, 204, 211, 213.
Apollinaris, grandfather of Sidonius, 109.
Apollinaris, son of Sidonius, 4, 84, 85, 162, 178.
Aprunculus, bishop of Langres, 180, 211.
Apuleius, his 'quaestiones', 72.
Aquilinus, 10, 11 n.[3].
Arabundus (*vide infra*, Arvandus), 103 n.[3].
Arbitrators, mentioned by Sidonius, 120.
Arianism of Theodoric II, 23, 24; of Euric, 92, 154.
Aristotle, studied in Gallic schools, 6, 7, 13.
Arithmetic, studied in Gallic schools, 6, 7.
Arles [Arelate], 10, 11, 28, 29, 38, 50, 52–7, 66, 67, 79, 125, 127, 149, 150, 152, 159, 164, 203, 209.
Arvandus, praetorian prefect of Gaul (A.D. 464–8), viii, 84, 103–7, 112, 166.
Asceticism, 70.
Astrology, 7.
Astyrius, consul (A.D. 449), 1 n.[3], 10.
Athenaeum, 7 n.[1].
Athenius, 55.
Aturres [Aire (Landes)], residence of Visigothic kings, 113 n.[3].
Audollent, A., xi.
Augustine, 15, 86, 118.
Ausonius, 3, 4, 14 n.[1], 74.
Auvergne, 9, 52, 60, 64, 87, 100 n.[6], 108, 111, 112, 139–41, 145, 149, 150, 152–4, 157–61, 179, 182–6, 197–200, 204, 206, 208.

Avienus, Gennadius, 97.
Avitacum, Sidonius' villa, 20, 63, 68, 71, 72 n.[4], 85; site of, 185–96.
Avitianus, *comes*, 115.
Avitus, bishop of Vienne, 18, 162.
Avitus, Eparchius, western emperor (A.D. 455–6), 9, 19, 22, 23, 26–31, 33, 36–41, 44, 58, 72, 99, 163, 166.
Avitus, schoolfellow of Sidonius, 10 n.[5], 149, 150, 201, 204.
Bacaudae, 2, 3 n.[2].
Barbarians, attitude of Sidonius towards, 48, 49, 66; food of, 71 n.[1]; *hospites*, 74 n.[9], 113, 163, 164.
Basilius, bishop of Aix-en-Provence, 158, 159, 207–10.
Basilius, Caecina, Roman senator, 97.
Baths, 72.
Bayeux, 77.
Baynes, Prof. N., xi.
Bayonne, lobsters from, 69.
Beer, 71.
Bible, read by Gallic nobles, 5, 73.
Bilimer, *magister militum Galliarum*, 151.
Bishop, laymen appointed as, 114; authority of, 115–20, 134; election of, 122–9.
Bordeaux, 27, 65, 66, 142, 164; oysters from, 69.
Bourg-de-Déols [Dolensis Vicus (Indre-et-Loire)], 65, 67.
Bourg-sur-Gironde [Burgus Pontii Leontii (Gironde)], 65, 67.
Bourges, 112; episcopal election at, 2, 12, 114, 126–9, 141, 206, 207.
Bréhier, E., xi.
Bretons, 94, 104, 139.
Brice, bishop of Orléans, 124.
Brigandage, 76.
Brittany (or Britain), journey to, 77.
Burgundians, 21, 25, 26, 41, 42, 45, 46, 66, 78, 91 n.[3], 94, 104, 112, 139, 140, 145, 148, 150, 157, 164, 181–5, 201, 205, 207 n.[6], 208 n.[3], 211 n.[1].
Burgundio, 6.
Caecina Basilius, see Basilius.
Calminius, 146 n.[5].
Camillus, 54.
Catullinus, 53, 66, 182.
Celestine, Pope (A.D. 422–32), 123, 125.
Celtic, revival of, 82.
Chalon-sur-Saône [Cabillonum], episcopal election at, 126–9, 131.
Chariobaudus, 70.
Chilperic, king of Burgundians, 157.
Church, as career, 78, 79; wealth of, 121, 122; attitude towards barbarians, 154.
Cicero, studied in Gallic schools, 4, 14.
Claudian, 5, 14 n.[1], 31, 32, 35, 87.
Claudianus Mamertus, brother of Mamertus (*vide infra*), 7, 11, 12, 15, 17, 18, 135, 138, 172 n.[3].

Date Due